Achieving TABE® Success in Mathematics

Level D

Senior Editor: Diane Nieker
Executive Editor: Linda Kwil
Marketing Manager: Sean Klunder
Production Manager: Genevieve Kelley
Cover Designer: Vickie Tripp

ISBN: 0-07-704469-X

Send all inquiries to:
Wright Group/McGraw-Hill
One Prudential Plaza
130 East Randolph Street, Suite 400
Chicago, IL 60601

Printed in the United States of America.

16 17 MAL 15 14

The McGraw·Hill Companies

Table of Contents

To the Learner

In our global technological society, mathematics skills are more important than ever. The pages in *Achieving TABE Success in Math, Level D* provide instruction to help you increase your understanding of many basic mathematical ideas. These pages also include problems that allow you to practice your skills and strengthen that understanding.

Start by taking the pretest on pages 1–6. Then use the answer keys on page 7 to check your answers. The evaluation charts for each set of answers identify the skill area for each problem in the pretest. These charts can help you pinpoint your areas of strength and determine any skill areas that need attention. Use the text pages listed for problems you miss to get additional information and practice.

The main part of the book is organized into sections by skill area: The Number System—Whole Numbers; Computation with Whole Numbers; Problem Solving; Decimals; Fractions; Integers; Ratio, Proportion, and Percent; Algebra; Geometry; Measurement; and Probability, Data, and Statistics. As you complete each page, check your answers using the answer key at the back of the book. Rework problems that you miss, and be sure to check with your instructor if you find you are having difficulty. At the end of each section you will find a Skills Checkup that serves as a quiz for that section. Answers for each Skills Checkup can be found in the answer key at the back of the book.

The posttest on pages 166–175 will measure your overall progress and determine whether you have mastered the skills you need to move to the next level. By successfully completing *Achieving TABE Success in Math, Level D*, you will have a strong foundation on which to continue building your skills in mathematics.

Correlation Chart

Correlations Between Contemporary's Instructional Materials and TABE® Math Computation and Applied Math

Test 2 Mathematics Computation

Subskill	TABE® Form 9	TABE® Form 10	TABE® Survey 9	TABE® Survey 10	Practice and Instruction Pages			
					Achieving TABE Success in Mathematics, Level D	*Pre-GED Mathematics*	*Complete Pre-GED*	*The GED Math Problem Solver**
Multiplication of Whole Numbers								
regrouping: by 1 digit	3, 7	2, 4	1, 5	1, 2	19	18 82–83	663–670	64–66, 69–70,
regrouping: by 2+ digits	5, 11, 15	6, 20, 22	3, 10	4, 14	21	18	663–670	64–66, 69–70, 82–83
Division of Whole Numbers								
no remainder: by 1 digit	6, 16	17	2, 11	13	24–25	19–21	671–673	64–66, 69–70
no remainder: by 2+ digits	—	27	—	18	28–29	19–21	671–673	64–66, 69–70
remainder: by 1 digit	4, 8, 22	7, 19, 23	4, 17	5, 15	26, 29	19–21	671–673	64–66, 69–70
Decimals								
addition	1, 12	1, 15	7	12	43	57–61, 64–67	716–717	189–195
subtraction	14, 26	3, 13	9	10	44	57–61, 64–67	717	189–195
multiplication	9, 13	8, 26	8	19	45–46	57–61, 67–69	718	189–195
division	2, 20	11, 14	15	8	47–49	57–61, 70–72	719	189–195
Fractions								
addition	28, 31	21, 31	20	21	60, 64	78–100	729–735	162–172
subtraction	10, 24	5, 24	6	3	61, 64	78–100	729–735	166–172
multiplication	17, 34	16, 29	12	11	65–66	78–90, 101–102	736–739	176–180
division	21, 33	33, 38	16	22	67–69	78–90, 102–104	740–743	182
Integers								
addition	25, 29	9, 18	18	6	75			48–52
subtraction	37, 39	36, 39	24	24	76			48–52
multiplication	19, 27, 36	10, 12, 32	14, 22	7, 9	77			96–97
division	18, 35	25, 34	13	16	77			96–97
Percents								
percents	23, 30, 32, 38, 40	28, 30, 35, 37, 40	19, 21, 23, 25	17, 20, 23, 25	86–96	125–143	753–766	226–239, 242–253

Test 3 Applied Mathematics

Subskill	TABE® Form 9	TABE® Form 10	TABE® Survey 9	TABE® Survey 10	Practice and Instruction Pages *Achieving TABE Success in Mathematics, Level D*	*Pre-GED Mathematics*	*Complete Pre-GED*	*The GED Math Problem Solver**
Number and Number Operations								
read, recognize numbers	4	—	—	—	8–9, 40	12–15	655–656	
compare, order	—	19	—	—	10, 40, 63	82–83, 88–89 153–157, 188–189	727–729	20–21,
place value	6	31	—	—	8, 40	10–11	653–654	
fractional part	39	49	19	24	53–54, 56	78–81	723–727	150–157
operation properties	7	44	4	21	100–101	38–39, 193–194		100–101, 104, 117–119
equivalent forms	9	17	5	—	55–57, 61	82–87	753–757	153–157, 162
number line	1	1	—	1	73–74, 79, 129–130			19, 48–52
factors, multiples, divisibility	—	47	—	22	58–59, 62	15–21, 64–72	657–673	69–71
percent	2	—	2	—	88–96	125–134, 136–140	753–766	226–239, 242–253
ratio, proportion	16	37	—	—	81–85	112–122, 136–140	745–752, 764–765	202–203, 207–211, 214–219, 231–232
Computation in Context								
whole numbers	28	10, 24	—	—	35–36	16–20, 38–43, 45–49	658, 668, 672, 675	** 28–31, 210–211, 237
decimals	19, 43	7	22	7	50, 96	34–37, 50–51, 72–77	720	** 90–97, 107–109, 190
fractions	24	36, 48	—	23	69–70, 96	96, 100, 110–111	735, 737, 742–743	** 90–97, 172, 188, 193, 218–219

Subskill	TABE® Form 9	TABE® Form 10	TABE® Survey 9	TABE® Survey 10	Practice and Instruction Pages *Achieving TABE Success in Mathematics, Level D*	*Pre-GED Mathematics*	*Complete Pre-GED*	*The GED Math Problem Solver**
Estimation								
estimation	15, 50	9, 15	25	8	11, 17, 22, 27, 37, 132–135	23–25, 44–45	678–679, 715–716	16–17, 59–60, 94–95, 170–171, 181, 242–243
rounding	5, 14, 38	5, 46	—	—	12, 40	62–63	676–677, 713–714	72–74, 85
reasonableness of answer	—	16	—	—	32, 132–133, 135	34–43	697–698	8–9, 72–73
Measurement								
appropriate unit	30	23	—	—	157	212–221	801–805	77
time	11, 29	18	6	—	138–140			
perimeter	41	14	20	—	141–142	225–228	808–809	13–14, 105
area	40	28	—	12	143–144	229–233	810–811	78–83, 116, 130–131
rate	—	32	—	15	86–87	115–117	747–748	203–204, 206–207, 223
convert measurement units	12	—	7	—	134, 136–138	212–218	801–803, 805	77, 169, 180
angle measure	—	34	—	17	118–119	222–224	807, 813–815	38–41, 44–45
Geometry and Spatial Sense								
symmetry	—	45	—	—				
plane figure	33	35	14	18	116–120	222–233	808–811	78–83
solid figure	—	27	—	11	122–123	233–235	812–813	84
visualization, spatial reasoning	—	30	—	—	121			
parallel, perpendicular	—	29	—	13	115	220–224	806	44
triangles	8	—	—	—	118	220–224	815–816	42–43, 220–221
angles	—	38	—	—	114	220–224	807, 813–815	38–41, 44–45
coordinate geometry	48	—	24	—	126	204–207		52–55, 254–257
parts of a circle	3	—	—	—	120	225	808	126–128
point, ray, line, plane	10	—	1	—	113	220–224	806	
transformations	34	—	—	—	124–125			

Subskill	TABE® Form 9	TABE® Form 10	TABE® Survey 9	TABE® Survey 10	Practice and Instruction Pages			
					Achieving TABE Success in Mathematics, Level D	*Pre-GED Mathematics*	*Complete Pre-GED*	*The GED Math Problem Solver**
Data Analysis								
bar, line, and circle graph	20, 35, 36	3, 4	10, 15, 16	3, 4	159–160, 162–163	149–172	772–777	10, 20, 140, 178, 236, 251, 258
table, chart, diagram	27	6	—	6	154–157		769–771	140, 173, 187, 205, 219, 236, 251, 258
conclusions from data	21	25	11	—	157, 160, 163	149–172	772–777	10, 140, 178, 187, 205, 219, 251, 258
appropriate data display	18, 26	13, 39	—	—	157	149–172	772–777	140, 236, 258
Statistics and Probability								
probability	17, 25	21	—	—	151–152	173–180	778–782	158, 179
statistics	37	22, 41	17	19	158	181–184 205, 219, 251		20, 173, 187,
sampling	44	50	23	25	153	181–184		
Patterns, Functions, Algebra								
number pattern	—	2	—	—	99			
variable, expression, equation	13, 32	40	8, 13	—	103–105, 109	196–199	785–792	14–15
function	45, 46	20	—	14	102			
inequality	22	33	9	16	105, 108–109		712, 727–729	
linear equation	47	26, 42	—	20	106–107, 109–110	196–203	789–798	27–33, 92–93, 259–261
Problem Solving and Reasoning								
solve problem	31	12	12	10	32–37, 70, 78	34–43, 45–49	689–696	2–7, 22, 59–60, 108–109
identify missing/ extra information	23, 49	11, 43	—	9	34	45–49	691–692	61, 144–145
evaluate solution	42	8	21	—	32–37	34–49	697–702	8–9, 141

* *The Math Problem Solver* can be substituted for *The GED Math Problem Solver*.

** All pages of "Check Your Understanding" and "GED Practice" contain computation in context problems.

Skills Inventory Pretest

Part A: Computation

Circle the letter for the correct answer for each problem.

1. $\begin{array}{r} 16 \\ \times\ 8 \\ \hline \end{array}$

   A 328
   B 200
   C 88
   D 128
   E None of these

2. $3\overline{)56}$

   F 18
   G 18 R2
   H 19
   J 19 R3
   K None of these

3. $2.45 − $1.96 =

   A $0.49
   B $1.49
   C $0.59
   D $0.48
   E None of these

4. $\begin{array}{r} 206 \\ \times\ 12 \\ \hline \end{array}$

   F 620
   G 2,492
   H 2,472
   J 2,462
   K None of these

5. $\frac{3}{15} + \frac{7}{15} =$

   A $\frac{2}{3}$ **C** $\frac{2}{5}$
   B $\frac{1}{3}$ **D** $\frac{1}{10}$
   E None of these

6. 25% of 160 =

   F 64
   G 25
   H 40
   J 400
   K None of these

7. $-10 + 2 =$

   A 8
   B -8
   C 12
   D -12
   E None of these

8. $-10 \times -5 =$

   F 50
   G -50
   H 15
   J 2
   K None of these

9. $\frac{3}{8} \times \frac{4}{9} =$

   A $\frac{7}{17}$ **C** $\frac{3}{8}$
   B $\frac{1}{6}$ **D** $\frac{4}{7}$
   E None of these

10. 6.37 + 0.043 =

    F 6.8
    G 0.6413
    H 6.3743
    J 6.413
    K None of these

11. $\begin{array}{r} 3\frac{1}{5} \\ -1\frac{1}{10} \\ \hline \end{array}$

    A $2\frac{1}{10}$ **C** $2\frac{1}{5}$
    B $2\frac{1}{15}$ **D** $2\frac{9}{10}$
    E None of these

12. $0.2503 \times 0.5 =$

    F 1.2515
    G 0.012515
    H 0.12515
    J 12.51
    K None of these

13. $13\overline{)273}$

A 20 R3
B 19 R25
C 21
D 19
E None of these

14. $6\overline{)3.42}$

F 0.057
G 0.57
H 57
J 0.507
K None of these

15. 40% of ☐ = 4

A 1
B 10
C 100
D 1,000
E None of these

16. What percent of 80 is 20?

F 40%
G 4%
H 25%
J 20%
K None of these

17. $\frac{1}{15} \div \frac{3}{5} =$

A $\frac{3}{5}$
B $\frac{3}{25}$
C $\frac{1}{9}$
D 1
E None of these

18. $-20 \times 8 =$

A 160
B -12
C -160
D 12
E None of these

19. $\frac{5}{6} + \frac{7}{12} =$

F $\frac{2}{3}$
G $1\frac{5}{12}$
H $1\frac{5}{6}$
J $2\frac{1}{6}$
K None of these

20. $9 - (-13) =$

A 22
B -4
C 4
D -22
E None of these

21. $12 - 3\frac{5}{7} =$

F $9\frac{5}{7}$
G $9\frac{2}{7}$
H $8\frac{2}{7}$
J $7\frac{2}{7}$
K None of these

22. $-50 \div -50 =$

A 1
B 2,500
C -1
D -2,500
E None of these

23. $8 + -4 + 5 =$

F 17
G -4
H 11
J -17
K None of these

24. $2.4 \times 3 =$

A 72
B 0.72
C 7.2
D 720
E None of these

25. $\begin{array}{r} 45 \\ \times\ 63 \\ \hline \end{array}$

F 2,535
G 2,035
H 2,735
J 2,835
K None of these

Part B: Applied Mathematics

Circle the letter for the correct answer to each question.

Byron created the graph below based on information he found about teens who said they had smoked during the past 30 days. Use the graph to answer Questions 1–4.

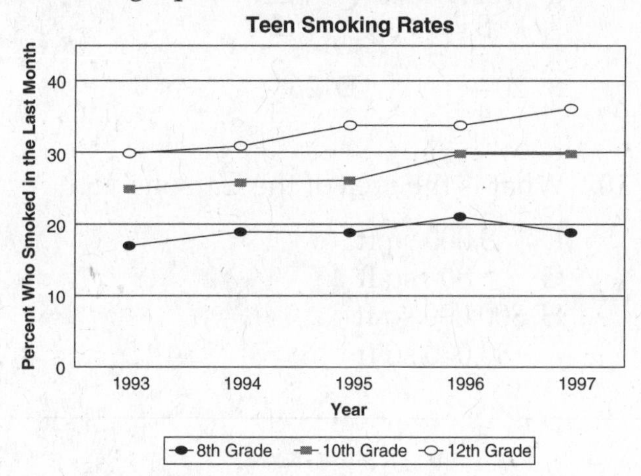

1. By about how much did the 12th-grade smoking rate rise between 1993 and 1997?

   A 12 percent
   B 5 percent
   C 18 percent
   D 30 percent

2. Which of the following best describes how the 8th-grade smoking rate changed during the period shown?

   F It rose steadily, without pause.
   G It never changed.
   H It rose slightly, but it has started to decline.
   J It has dropped drastically.

3. In what period was there the greatest increase in the percentage of 10th graders who smoked?

   A 1993 to 1994
   B 1994 to 1995
   C 1995 to 1996
   D 1996 to 1997

4. Which of the following statements is supported by the information on this graph?

   F As teenagers get older, they are more likely to smoke.
   G The older you are, the harder it is to stop smoking.
   H The older teenage smokers get, the more cigarettes they smoke per day.
   J The older teenagers get, the less likely they are to start smoking.

5. What is the value of n in the inequality in the box?

   $$n + 4 < 12$$

   A $n < 8$ **C** $n < 16$
   B $n > 16$ **D** $n > 8$

6. What is the measure of $\angle ACB$ in this triangle?

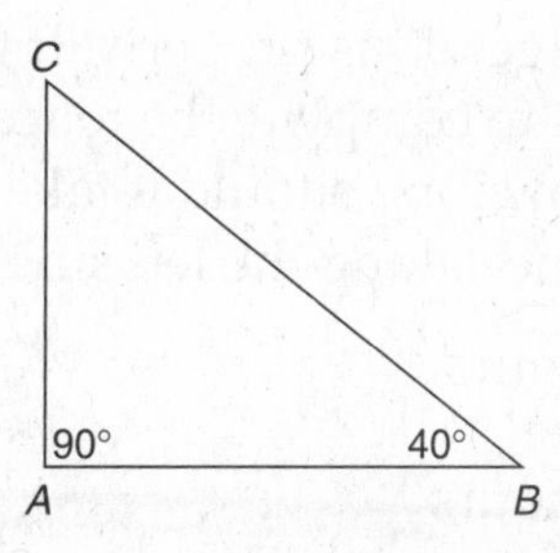

   F 45°
   G 40°
   H 50°
   J This cannot be determined.

The diagram below shows the layout of the Carsons' property. Use it to answer Questions 7–10.

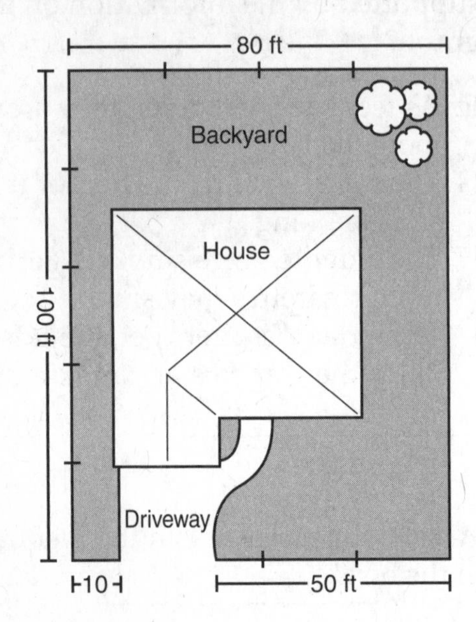

7. What is the perimeter of the Carsons' property?

   A 200 ft C 340 ft
   B 280 ft D 360 ft

8. Mr. Carson built a fence along the back edge of his property. It took him 5 hours to complete the fence. At that rate, how long should it take him to put up a fence along the left side of his lot?

   F 10 hours
   G $9\frac{1}{2}$ hours
   H 6 hours 15 minutes
   J 4 hours 20 minutes

9. Mr. Carson wants to string lights along the back fence. The lights are sold by the yard. How many yards of lights will he need to cover that length?

   A $10\frac{7}{8}$ C $30\frac{1}{2}$
   B $20\frac{1}{4}$ D $26\frac{2}{3}$

10. What is the area of the Carsons' lot?

    F 8,000 sq ft
    G 800 sq ft
    H 800,000 sq ft
    J 80,000 sq ft

11. Kayla's son is taking a train to Yorkville to visit his grandparents. His train is scheduled to leave the station at 3:45. If the train ride is 2 hours 20 minutes, what time should he arrive in Yorkville?

    A 5:05 C 6:05
    B 5:25 D 6:35

12. Which point on the number line indicates a number that is less than 5.3 and greater than 3.5?

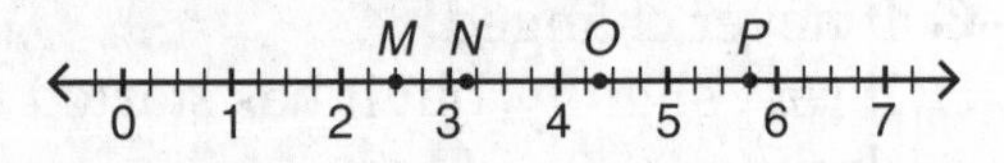

    F point M
    G point N
    H point O
    J point P

Use the figure shown below to answer Questions 13–15. Each vertex on the figure has been labeled, and the length of each side is shown.

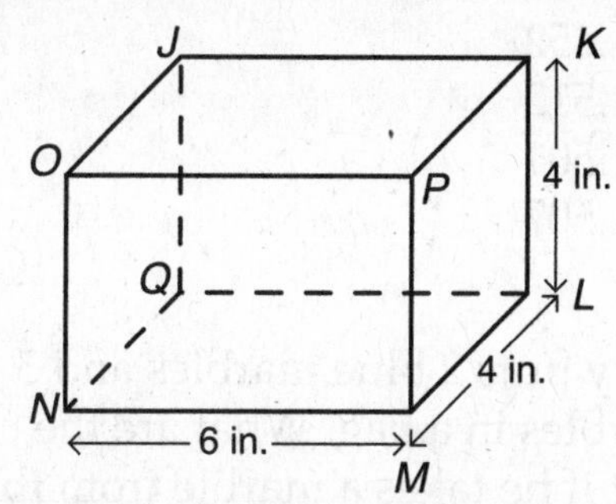

13. How many faces does the figure have?

A 4 C 8
B 6 D 10

14. Which line segments of the figure are perpendicular?

F $\overline{JK}$ and $\overline{MN}$ H $\overline{JO}$ and $\overline{KP}$
G $\overline{JQ}$ and $\overline{KL}$ J $\overline{JK}$ and $\overline{JO}$

15. What is the total surface area of the figure?

A 112 sq in. C 128 sq in.
B 124 sq in. D 284 sq in.

16. Which of these patterns could be folded to make the figure below?

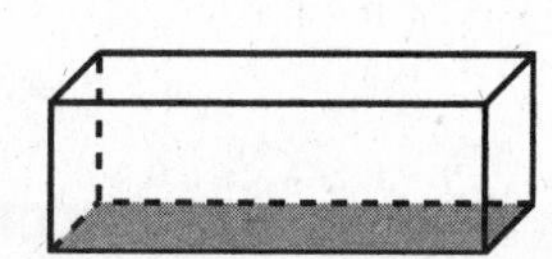

F H

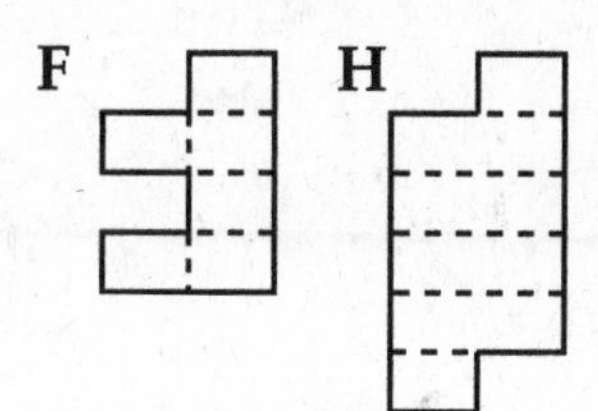

G J

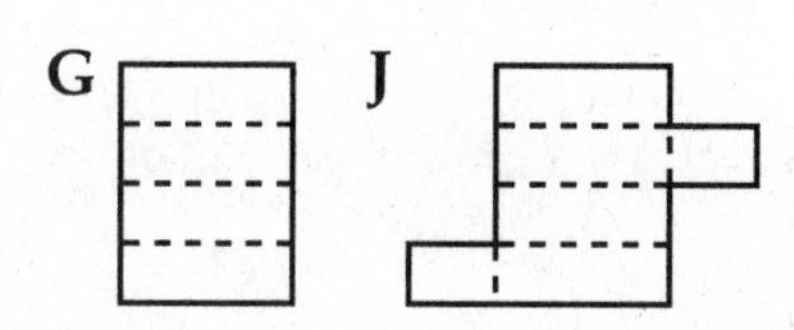

Use this chart for Numbers 17–20.

Fried Rice Mix Nutrition Facts

Serving Size:	$\frac{1}{3}$ cup (44 grams)	
Servings per Container	5	
Amount Per Serving		
Calories	150	
Calories from Fat	10	
	% Daily Allowance	
Total Fat	1 g	2%
Saturated Fat	0 g	0%
Cholesterol	0 mg	0%
Sodium	600 mg	25%
Potassium	420 mg	12%
Total Carbohydrates	32 g	11%
Dietary Fiber	2 g	9%
Sugars	6 g	
Other Carbohydrates	24 g	
Protein	5 g	

17. This chart shows that 600 mg (milligrams) of sodium is 25% of the sodium you should eat each day. What is the recommended daily allowance of sodium?

A 1,200 mg C 2,400 mg
B 150 mg D 15,000 mg

18. How many cups of fried rice mix are in this container?

F 5 H $\frac{3}{5}$
G $1\frac{2}{3}$ J 3

19. About how many ounces of fried rice mix are in one serving? (1 ounce = 28 grams)

A 1.5 C 15
B 4 D 1,200

20. Which number sentence shows how many calories are in 1 cup of this mix?

F $150 \div 3 = \square$
G $150 \times 3 = \square$
H $3 \div 150 = \square$
J $150 = \square$

An end table is on sale for $123. To buy it, you must put down 15%. Use this information to do Numbers 21–23.

21. The table has been marked down 20%. Which of these is the best estimate of how much the table originally cost?

A $100
B $150
C $175
D $200

22. Which of these number sentences could you use to figure out how much is left to pay after you make the down payment?

F $\$123 \times 0.15 =$ ______

G $\frac{\$123}{6} =$ ______

H $\$123 - (\$123 \times 0.15) =$ ______

J $\$123 - 0.15 =$ ______

23. Mark and Cindy have $600 to buy furniture. If they buy an end table and a chair for $345.59, about how much will they have left? *(Round all numbers to the nearest 10 dollars.)*

A $250
B $130
C $ 70
D $160

24. Which of these has the same value when it is rounded to the nearest tenth and to the nearest hundredth?

F 1.452
G 1.512
H 1.501
J 1.507

25. Gary puts 2 blue marbles and 3 green marbles in a bag. What are the chances that if he takes a marble from the bag without looking, it will be blue?

A $\frac{2}{3}$

B $\frac{2}{5}$

C $\frac{1}{2}$

D $\frac{1}{5}$

Pretest Answer Key and Evaluation Chart

Use these answer keys to check your pretest. The evaluation charts match each problem in the pretest to a skill area. The charts will refer you to pages in this book that can provide information and practice to help you with problems you missed.

Answer Key—Part A: Computation

1.	D	14.	G
2.	G	15.	B
3.	A	16.	H
4.	H	17.	C
5.	A	18.	C
6.	H	19.	G
7.	B	20.	A
8.	F	21.	H
9.	B	22.	A
10.	J	23.	K
11.	A	24.	C
12.	H	25.	J
13.	C		

Evaluation Chart—Part A: Computation

Problem Number	Skill Area	Text Pages
1, 4, 25	Multiplication of Whole Numbers	18–22 and 30–31
2, 13	Division of Whole Numbers	23–29 and 30–31
3, 10, 12, 14, 24	Decimals	40–52
5, 9, 11, 17, 19, 21	Fractions	53–72
7, 8, 18, 20, 22, 23	Integers	73–80
6, 15, 16	Percents	88–98

Answer Key—Part B: Applied Mathematics

1.	B	14.	J
2.	H	15.	C
3.	C	16.	J
4.	F	17.	C
5.	A	18.	G
6.	H	19.	A
7.	D	20.	G
8.	H	21.	B
9.	D	22.	H
10.	F	23.	A
11.	C	24.	H
12.	H	25.	B
13.	B		

Evaluation Chart—Part B: Applied Mathematics

Problem Number	Skill Area	Text Pages
12, 22	Number and Number Operations	30–31, 40–49, 53–69, 73–78
17, 18, 19	Computation in Context	50, 70, 87, 94–96
23, 24	Estimation	17, 22, 27, 37, 42
6, 7, 9, 11	Measurement	129–150
10, 13, 14, 15, 16	Geometry and Spatial Sense	113–128
1, 2, 3, 4	Data Analysis	155–157, 159–163, 164–165
25	Statistics and Probability	151–153, 158, 164–165
5, 20	Patterns, Functions, Algebra	99–112
8, 21	Problem Solving and Reasoning	32–39

The Number System – Whole Numbers

Reviewing Place Value

Ten **digits** — 0, 1, 2, 3, 4, 5, 6, 7, 8, and 9 — are used in our number system. The value of a digit in a number depends on its **place** in that number. For example, 17 and 71 have the same digits, but because the digits are in different places, they are different numbers.

17 = 10 + 7 or 1 ten and 7 ones
71 = 70 + 1 or 7 tens and 1 one

millions	hundred thousands	ten thousands	thousands	hundreds	tens	ones
___	___	9	7,	4	0	3

The diagram at the right shows the names of the first seven place-value columns for whole numbers. The number 97,403 has digits in the first five places. It has 9 ten thousands, 7 thousands, 4 hundreds, 0 tens, and 3 ones. It has no digit in the millions place or in the hundred thousands place.

PRACTICE

Fill in the blanks below.

1. 35,017 has 3 __________ , 5 __________ , 0 __________ , 1 __________ , and 7 __________.
2. 107,403 has ____.
3. 1,500,000 has ____.
4. In what place is 6 in the number 3,962,710? __________
5. In what place is 5 in the number 4,510,329? __________
6. What digit is in the thousands place in the number 92,613? __________
7. What digit is in the hundreds place in the number 3,167,503? __________
8. What is the value of the 5 in 5,432,106? __________
9. What is the value of the 6 in 4,605,912? __________
10. What is the value of the 7 in 117,500? __________
11. What is the value of the 3 in 310,416? __________
12. What is the value of the 5 in 6,750,000? __________

Naming Large Numbers

The digits in our numbers are grouped in families: ones family, thousands family, millions family, billions, trillions, and so on.

	millions			thousands			ones			
...	hundred millions	ten millions	ones	hundred thousands	ten thousands	thousands	hundreds	tens	ones	...
					4	1	0	3	2	
		3	4	5	0	2	4	7	0	

Each family has 3 members: ones, tens, and hundreds. The 3 digits are read as a unit followed by the family name. (with the exception of the ones family name, which is not said.) Commas separate the families and make it easier to read the number.

Examples

41,032 = 40,000 + 1,000 + 30 + 2
41,032 is read *41 thousand, 32.*
Written in words, 41,032 is *forty-one thousand, thirty-two.*
Note: Places that contain a zero are not named.
For example, the word *hundreds* is not included in 41,032.

34,502,470 = 30,000,000 + 4,000,000 + 500,000 + 2,000 + 400 + 70
34,502,470 is read *34 million, 502 thousand, 470.*
Written in words, 34,502,470 is *thirty-four million, five hundred two thousand, four hundred seventy.*

PRACTICE

Write each number below in words.

1. 59,505 ____________________
2. 211,525 ____________________
3. 4,500,072 ____________________
4. 901,316 ____________________
5. 23,012 ____________________

Write each number in digits. Watch for number names that skip places and write zeros in those places.

6. ninety-one thousand, two hundred one ________
7. sixty-three thousand, four hundred twelve ________
8. three million, four hundred thousand ________
9. eight hundred one thousand, three hundred fifty-six ________

Comparing Whole Numbers

- If two whole numbers have a different number of digits, the number with more digits is greater.

Example
5,013 is greater than 984 because it has more digits.

- If two whole numbers have the same number of digits, start at the left to compare them.

Example
5,862 is greater than 2,974 because 5 thousand is greater than 2 thousand.

PRACTICE

Circle the greater number in each problem.

1. 12,301 946
2. 103,812 946,203
3. 55,000 9,415
4. 92 112
5. 80,912 91,415
6. 695,415 1,000,000
7. 12,000 9,946
8. 45,678 45,078
9. 1,512 1,510
10. 999,999 100,000

Circle the greatest number in each problem.

11. 712 1,042 900
12. 888 899 875
13. 90 800 812
14. 750 92 623
15. 511 595 1,596
16. 645 654 564

Arrange each set of numbers from least to greatest.

17. 397 115 269 52
18. 6,710 952 1,736 9,000
19. 17,008 5,107 5,123 952 978

Arrange each set of numbers from greatest to least.

20. 99 1,000 425 8,056
21. 3,280 997 4,000 3,985
22. 1,480 1,048 1,408 1,840

Arrange these digits to make the greatest number possible.

23. 2 0 1 9
24. 5 8 1 6
25. 1 3 7 2

Arrange these digits to make the least possible number.

26. 5 0 2 3
27. 6 8 7 1
28. 3 4 0 2

Using Estimation

Often, you do not need to know an exact amount. You just need to know *about* how much. *About how much will it cost? Approximately how far is it? Around how much time will it take?* In such cases, an estimate is enough. An **estimate** is a number that is close to the actual amount.

PRACTICE

Circle the number of each question or situation that can be answered with an estimate.

1. Hiram needs to figure the balance in his checking account.
2. Reagan is having a barbecue and has invited several friends. About how much potato salad will he need to buy?
3. It takes Carrie almost two hours to do her laundry. If she starts at 3 o'clock, about what time will she be done?
4. You want to know if painter A will charge more to paint your house than painter B.
5. A co-worker says that 150 people work in your office. You want to know whether that number is correct.
6. About how many cookies are needed for the bake sale?

Estimating is an important skill—one that you should get into the habit of using. An estimate can give you an idea of about what your answer should be before you begin to calculate. You can also use estimates to check your work after you have done your calculations.

You can often use common sense to estimate whether an answer is correct. For instance, you know from experience that a hamburger and fries should cost about $5, not $50 and not 50 cents.

PRACTICE

Use common sense estimation to answer each question below.

7. Jenny's lunch cost $13.75. She paid with a $20 bill. About how much change should she receive?

   A 60 cents　　**C** 60 dollars
   B 6 dollars　　**D** 6 cents

8. About how much should 20 gallons of gasoline cost?

   F $5　　**H** $50
   G $1.50　　**J** $200

9. About how much time would you need for a 500-mile drive on a highway?

   A 9 hours　　**C** 27 hours
   B 3 hours　　**D** 3 days

10. If you earn $15 an hour, about how much would you earn in one day?

    F $300　　**H** $100
    G $ 25　　**J** $800

11. About how much is the sales tax on a purchase of $15?

    A $8　　**C** 3¢
    B $1.25　　**D** $10

12. About how much would a pound of boiled ham cost?

    F 25 cents　　**H** 3 dollars
    G 15 cents　　**J** 8 dollars

Rounding

Another way to make an estimate is to calculate with rounded numbers. A **rounded number** is close to the exact one, but it is easier to work with. Rounded numbers end with one or more zeros.

When rounding a number, you might imagine the number sitting on part of a hilly number line like the example shown here for rounding to the nearest ten. When you round to the nearest ten, if the digit in the ones place is 0, 1, 2, 3, or 4, the number rolls back to the nearest low spot. If the digit in the ones place is 5, 6, 7, 8, or 9, the number rolls ahead to the nearest low spot.

107 108 109 110 111 112 113 114 115 116 117 118 119 120 121 122 123

Examples
111 rounds to 110
118 rounds to 120
135 rounds to 140

You can round the same number to different place values. Look at the digit *to the right* of the place you are rounding to. If that digit is less than 5, roll back. If that digit is 5 or greater, roll ahead.

Examples

2,186 rounded to the nearest ten is 2,190
rounded to the nearest hundred is 2,200
rounded to the nearest thousand is 2,000

PRACTICE

Circle the amount that shows the number rounded to the nearest ten.

1.	182 rounds to	180	190
2.	807 rounds to	800	810
3.	99 rounds to	90	100
4.	4 rounds to	0	10

Circle the amount that shows the number rounded to the nearest hundred.

5.	439 rounds to	400	500
6.	1,249 rounds to	1,200	1,300
7.	49 rounds to	0	100
8.	972 rounds to	900	1,000

Circle the amount that shows the number rounded to the nearest dollar.

9.	$45.67	$45.00	$46.00
10.	$19.95	$19.00	$20.00
11.	$ 0.75	$0.00	$1.00
12.	$128.37	$128.00	$129.00

Round each number as indicated.

13. 3,053 to the hundreds place ________

14. 12,031 to the thousands place ________

15. $3.03 to the nearest dollar ________

16. $12.56 to the nearest dime ________

17. $89.75 to the nearest 10 dollars ________

18. 570 to the nearest thousand ________

Number System Skills Checkup

Circle the letter for the correct solution to each problem.

1. Which of these is another way to write 7,000 + 500 + 30?

   A 753
   B 7,530
   C 7,503
   D 70,530

2. Which of these dollar amounts is greater than $1.50 but less than $1.75?

   F $1.25
   G $1.92
   H $1.63
   J $1.45

3. Round the number 135,789 to the nearest ten thousand.

   A 130,000
   B 135,790
   C 136,000
   D 140,000

4. What is the total value of three quarters, two dimes, and five nickels?

   F $0.95
   G $0.85
   H $1.05
   J $1.20

5. Which of these numbers is fifty-four thousand, sixty?

   A 54,006
   B 54,060
   C 5,460
   D 5,406

6. Which of these is the correct way to write one thousand twelve dollars?

   F $1,120 H $1,012
   G $1,102 J $1,201

7. Which choice shows the numbers in order from least to greatest?

   A 25,146 25,832 24,964 25,819
   B 24,964 25,146 25,832 25,819
   C 24,964 25,146 25,819 25,832
   D 25,146 25,819 25,832 24,964

8. Which group of coins is worth exactly $\frac{1}{2}$ of a dollar?

   F two dimes and a quarter
   G a quarter and four nickels
   H three nickels and three dimes
   J a quarter and five nickels

9. Which of these is the most reasonable estimate of how much it would cost to buy T-shirts for all 20 members of a soccer team?

   A $ 20
   B $ 200
   C $ 2,000
   D $20,000

10. In which number is the digit in the hundred thousands place greater than that of the number in the box?

5,843,207

    F 2,754,163
    G 7,295,086
    H 4,968,637
    J 6,472,405

Number System Skills Checkup (continued)

Crystal is an assistant manager at a department store. Part of her job is to keep track of the store's sales. This list shows the sales figures for yesterday. Use the list to answer Questions 11–16.

Yesterday's Sales Figures	
Housewares	$9,475.15
Hardware	$12,809.98
Clothing	$10,870.56
Office Supplies	$4,750.42

11. Which department took in the most money yesterday?

   A Housewares
   B Hardware
   C Clothing
   D Office Supplies

12. To the nearest one thousand dollars, how much money did the clothing department take in?

   F $9,000 H $11,000
   G $10,000 J $12,000

13. Crystal's boss asked her to round the sales figures to the nearest one hundred dollars. Which of these lists shows yesterday's sales rounded to the nearest one hundred dollars?

   A $9,500, $13,000, $11,000, $5,000
   B $9,500, $12,800, $10,800, $4,800
   C $9,500, $12,800, $11,000, $4,000
   D $9,500, $12,800, $10,900, $4,800

14. Which of these is the most reasonable estimate of the store's total sales yesterday?

   F $35,000 H $15,000
   G $10,000 J $100,000

15. Crystal calculated that last year's total sales were $19,560,000. What is this amount in words?

   A nineteen million, five hundred sixty dollars
   B nineteen million, five hundred sixty thousand dollars
   C nineteen billion, five hundred sixty million dollars
   D nineteen million, five hundred six thousand dollars

16. For which of the following tasks could Crystal most reasonably use an estimate?

   F making out a paycheck
   G putting prices on blow dryers
   H filling out her time card
   J reporting how many customers came to the store last year

17. Which two numbers will both round to 500?

   A 492 and 435 C 550 and 450
   B 515 and 565 D 480 and 520

18. Which of these is the most reasonable estimate of how long it would take an adult to walk 5 miles?

   F 5 minutes H 2 hours
   G 30 minutes J 10 hours

19. How many of the amounts shown in the box round to $42.20?

$42.15 $42.12 $42.19 $42.09

   A 0 C 2
   B 1 D 3

Computation with Whole Numbers

Reviewing Addition

When you add, you put amounts together to find a **sum** or **total**. Here are some important things to keep in mind about addition.

- When you add zero to a number, the sum is the number. $6 + 0 = 6$
- In addition, you can change the order of the **addends** (the numbers being added) and the sum or total will *not* change. $7 + 4 = 11$, and $4 + 7 = 11$
- You can add only two numbers at a time. In addition, you can change the way you group the numbers and the answer will not change.

$$(3 + 4) + 5 = 3 + (4 + 5)$$
$$7 + 5 = 3 + 9$$
$$12 = 12$$

Because the order does not matter, you can start with any two digits and add pairs that are easier for you to work with. Look at the example to the right.

$$8 + 6 + 5 + 2$$
$$10 + 6 + 5$$
$$15 + 6 = 21$$

- You must add digits that have the same place value—add ones to ones, tens to tens, hundreds to hundreds, and so on. Start with the ones column and work to the left, adding one column at a time. Regroup as needed.

```
   1 2
 2,497
    17
+  309
 2,823
```

- You can use subtraction to check addition. You put two parts together to find a total. If you subtract one of the parts from that total, you will have the other part.

```
  25      42      42
+ 17    - 17    - 25
  42      25      17
```

PRACTICE

Find each sum. If the addends have labels or dollar signs and decimal points, be sure to include them in your answer. Use subtraction to check your answers.

1. 711 + 350
2. 80,215 + 4,552
3. 621 gal + 52 gal
4. 517 + 49
5. 12,402 + 499 = ________
6. \$12.16 + \$6.57 = ________
7. 45 min + 75 min = ________
8. 5,324 + 78 = ________
9. 107 + 89 + 1,230 = ________
10. 15,054 + 2,006 = ________
11. 213,456 + 875 = ________
12. 75 + 1,092 + 350 = ________
13. 69 + 1,678 = ________
14. 1,100 + 173 = ________
15. 47 + 365 + 9 = ________
16. 95 + 268 = ________

Reviewing Subtraction

When you subtract, you compare amounts, or, you remove a part from the total. The result is called the **difference**. Here are some important things to keep in mind about subtraction.

- When you subtract zero from a number, the difference is the number.
  $6 - 0 = 6$

- If you change the order of the numbers you are subtracting, the difference will *not* be the same. $9 - 3 = 6$, but $3 - 9 \neq 6$

- You can subtract only two numbers at a time. Unlike addition, in subtraction, if you change the way you group the numbers, you will get a different answer.

$$(8 - 5) - 3 \stackrel{?}{=} 8 - (5 - 3)$$
$$3 - 3 \stackrel{?}{=} 8 - 2$$
$$0 \neq 6$$

- You must subtract digits that have the same place value—subtract ones from ones, tens from tens, hundreds from hundreds, and so on. Start with the ones column and work to the left, subtracting one column at a time. Regroup as needed.

$$\begin{array}{r} 4{,}500 \\ -\ 2{,}917 \\ \hline 1{,}583 \end{array}$$

- Addition can undo subtraction. When you subtract one part from the total, you will be left with a second part. If you put the parts back together, you will have the total you started with. You can use addition to check subtraction.

$$\begin{array}{r} 31 \\ -\ 21 \\ \hline 16 \end{array} \qquad \begin{array}{r} 21 \\ +\ 16 \\ \hline 37 \end{array}$$

PRACTICE

Find each difference. If the numbers have labels or dollar signs and decimal points, be sure to include them in your answer. Use addition to check your answers.

1. 333 miles − 110 miles

2. 1,670 − 420

3. 3,710 − 590 = ________

4. $40 − $2.50 = ________

5. 596 − 59 = ________

6. 2,500 − 250 = ________

7. 201 − 85 = ________

8. $42.49 − $30.25

9. 1,590 − 1,457

10. 2,507 − 683 = ________

11. 15,347 − 3,849 = ________

12. 3,006 − 148 = ________

13. 50,000 − 2,016 = ________

14. 400 − 374 = ________

Estimating Sums and Differences

Estimation is an important math tool that can help you find *about* what the exact answer will be before you begin computing. You can also use estimation after doing calculations as a way to check to see if your answer is reasonable. You can do much of the estimating in your head. The symbol ≈ means "is approximately equal to" and is used to show that an answer is an estimate.

One method of estimating is to **round** each number *before* you compute. You can round the numbers to different place values. Depending on how you round, the answers may not be exactly the same, but they should be close to the actual answer. Remember, when you estimate, you are looking for an approximate answer.

Examples Estimate the sum of 657 and 1,920.

To nearest hundred

657 ≈ 700 and 1,920 ≈ 1,900
700 + 1,900 ≈ 2,600
657 + 1,920 ≈ 2,600

To greatest place value

657 ≈ 600 and 1,920 ≈ 2,000
600 + 2,000 ≈ 2,600
657 + 1,920 ≈ 2,600

Another way to estimate is to use **front-end estimation**. With this method, you write the front-end digit of each number and replace all of the other digits with zeros. Then compute using the rewritten amounts.

Example Estimate the sum of 657 and 1,920.

657 ≈ 600 and 1,920 ≈ 1,000
600 + 1,000 = 1,600
657 + 1,920 ≈ 1,600

Rounding generally gives a result closer to the actual answer than front-end estimation.

PRACTICE

Estimate each sum or difference. Pay careful attention to the signs. Show how you find each answer.

1. 746 + 219
2. 1,460 + 984
3. 23,846 – 2,863 ≈ ________
4. 19,450 + 5,471 ≈ ________
5. 622 + 1,745 ≈ ________
6. 92,447 + 4,896 ≈ ________
7. 34,186 – 13,850 ≈ ________
8. 57 − 9
9. 268 − 55
10. 446 + 3,812 ≈ ________
11. 5,275 + 173 ≈ ________
12. 594,720 – 2,366 ≈ ________
13. 75 + 185 + 1,336 ≈ ________
14. 9,567 − 3,269 ≈ ________

Reviewing Multiplication

Multiplication is a quick way to add when you have groups that are equal. For example, $30 + 30 + 30 + 30 = 120$ can also be written $4 \times 30 = 120$.
Here are some important things to keep in mind about multiplication.

$3 \times 5 = 15$ (3 and 5 are the factors; 15 is the product)

- The numbers that are multiplied are **factors**.
- The result of multiplying is called a **product**.

- When you multiply a number by one, the product is the number. $1 \times 358 = 358$

- When you multiply a number by zero, the product is zero. $0 \times 45 = 0$

- If you change the order of the factors you are multiplying, the product will *not* change.
  $3 \times 25 = 75$, and $25 \times 3 = 75$

- You can multiply only two numbers at a time. In multiplication, you can change the way you group the numbers and the answer will not change.

  $$(3 \times 4) \times 5 = 3 \times (4 \times 5)$$
  $$12 \times 5 = 3 \times 20$$
  $$60 = 60$$

  Because the order does not matter, you can start with any two factors. Find pairs that are easier for you to work with. In the above example, the grouping on the right might make it easier to find the product.

- Most multiplication problems are written in column form. Multiply each digit of the top number by the number on the bottom. Start with the ones column and work to the left. Multiply one column at a time.

```
  134
×   2 ← Multiply by this digit.
  268
```

2×4 (ones) $= 8$ (ones)
2×3 (tens) $= 6$ (tens)
2×1 (hundred) $= 2$ (hundreds)

PRACTICE

Find each product. Write problems in column form to multiply. Check your work.

1. 124×2 = ______
2. 310×3 = ______
3. 511×5 = ______
4. 81×6 = ______
5. $202 \times 4 =$ ______
6. $\$4.21 \times 2 =$ ______
7. $\$7.04 \times 2 =$ ______
8. $6{,}102 \times 4 =$ ______
9. $\$21 \times 8 =$ ______
10. $3 \times 523 =$ ______
11. $5 \times 60 =$ ______
12. $3 \times 213 =$ ______
13. $8 \times 211 =$ ______
14. $3 \times \$1.32 =$ ______
15. $402 \times 3 =$ ______
16. $\$3.10 \times 5 =$ ______

Regrouping in Multiplication

When you multiply, it is important to record the digits in the correct place-value columns. Products of 10 or greater are regrouped. This regrouping procedure, often called **carrying**, is similar to what is done in addition. You may have to **regroup**, or carry, more than one time in a multiplication problem. Study the example.

Example Multiply 278 by 6.

```
   4 4
   278
 ×   6
 1,668
```

1,668 ← 6 × 8 (ones) = 48 (ones)
Write 8 ones in the ones place.
Carry the 4 tens. Write 4 at the top of the tens column.

6 × 7 (tens) = 42 (tens)
42 tens + 4 tens* = 46 tens
Write 6 tens in the tens place.
Carry the 4 hundreds. Write 4 at the top of the hundreds column.

6 × 2 (hundreds) = 12 (hundreds)
12 hundreds + 4 hundreds* = 16 hundreds

**Note:* Multiply first, then add the number you carried from regrouping.

PRACTICE

Find each product. Write problems in column form. When you regroup, be sure to record any numbers you carry and add them to the product for that column.

1. 604 × 5
2. 89 × 7
3. 69 × 3
4. 5,009 × 6
5. 25 × 7
6. 775 × 3
7. 1,315 × 4
8. 3,625 × 2
9. 451 × 9
10. 40,361 × 8
11. 6 × 478 = ________
12. 902 × 9 = ________
13. 3 × 8,255 = ________
14. 572 × 7 = ________
15. 84 × 6 = ________
16. 8 × 3,890 = ________
17. 9 × 4,021 = ________
18. 3 × 2,730 = ________

Multiplying by 10s, 100s, and 1,000s

Multiplying a number by 10, 100, 1,000, and so on, is easy. You merely write the number and then write the corresponding number of zeros to complete the multiplication. Write one zero when multiplying by 10, two zeros when multiplying by 100, three zeros when multiplying by 1,000, and so on.

Examples $10 \times 5 = \underline{50}$ $\quad 100 \times 5 = \underline{500}$ $\quad 1{,}000 \times 5 = \underline{5{,}000}$

$10 \times 40 = \underline{400}$ $\quad 100 \times 40 = \underline{4{,}000}$ $\quad 1{,}000 \times 40 = \underline{40{,}000}$

Multiplying by multiples of 10 (20, 30, 40, 50, and so on) is also easy. Multiply the number by the digit in the tens place, and then write one zero.

Examples

20×5 — $2 \times 5 = 10$, then write a zero to get 100 — $20 \times 5 = \underline{100}$

30×40 — $3 \times 40 = 120$, then write a zero to get 1,200 — $30 \times 40 = \underline{1{,}200}$

When the problem is written vertically, write zeros in the answer before multiplying. Then multiply by the nonzero digit, working from left to right.

Examples

$$\begin{array}{r} 42 \\ \times\ 20 \\ \hline 0 \end{array} \qquad \begin{array}{r} 42 \\ \times\ 20 \\ \hline 840 \end{array} \qquad \begin{array}{r} 125 \\ \times\ 40 \\ \hline 0 \end{array} \qquad \begin{array}{r} {}^{1\,2} \\ 125 \\ \times\ 40 \\ \hline 5{,}000 \end{array}$$

PRACTICE

Find each product.

1. 10 × 384 = ________
2. 10 × 96 = ________
3. 40 × 93 = ________
4. 70 × 107 = ________
5. 100 × 96 = ________
6. 100 × 735 = ________
7. 300 × 25 = ________
8. 600 × 122 = ________
9. 20 × 47 = ________
10. 600 × 135 = ________
11. 542 × 30 = ________

12. 73 × 10
13. 35 × 10
14. 43 × 20
15. 15 × 30
16. 351 × 80
17. 10 × 37

18. 84 × 10
19. 96 × 10
20. 46 × 90
21. 235 × 50
22. 58 × 60
23. 87 × 70

24. 345 × 40
25. 91 × 20
26. 905 × 30
27. 270 × 800
28. 210 × 500
29. 450 × 300

Multiplying by a Two-Digit Number

When you multiply a number by a two-digit factor, you actually complete two separate multiplication problems to find **partial products**. Then you add the partial products.

Example
Find 637 × 35.

- Think of 35 as 30 + 5.
- Multiply 637 by 5.
- Multiply 637 by 30.
- Add the partial products to get the final product.

637 × 35 = 22,295

$$\begin{array}{r} 637 \\ \times\ 35 \\ \hline \end{array} \rightarrow \begin{array}{r} {}^{1\,3} \\ 637 \\ \times\quad 5 \\ \hline 3{,}185 \end{array} + \begin{array}{r} {}^{1\,2} \\ 637 \\ \times\ \ 30 \\ \hline 19{,}110 \end{array} \rightarrow \begin{array}{r} {}^{1} \\ 19{,}110 \\ +3{,}185 \\ \hline 22{,}295 \end{array}$$

partial products

You can do all of the steps as parts of one problem. Write the partial products one below the other and then add them.

$$\begin{array}{rl} {}^{1\,2} & \\ {}^{1\,3} & \\ 637 & \\ \times\ 35 & \\ \hline 3{,}185 & \leftarrow (5 \times 637) \\ +19{,}110 & \leftarrow (30 \times 637) \\ \hline 22{,}295 & \end{array}$$

PRACTICE

Complete each problem.

1. $\begin{array}{rl} {}^{5} & \\ 58 & \\ \times\ 47 & \\ \hline 406 & \leftarrow (7 \times 58) \\ + \qquad & \leftarrow (40 \times 58) \\ \hline \end{array}$

2. $\begin{array}{rl} {}^{1} & \\ 23 & \\ \times\ 86 & \\ \hline 138 & \leftarrow (6 \times 23) \\ + \qquad & \leftarrow (80 \times 23) \\ \hline \end{array}$

3. $\begin{array}{rl} 22 & \\ \times\ 55 & \\ \hline & \leftarrow (5 \times 22) \\ + \qquad & \leftarrow (50 \times 22) \\ \hline \end{array}$

4. $\begin{array}{r} 123 \\ \times\ \ 96 \\ \hline \end{array}$

5. $\begin{array}{r} 905 \\ \times\ \ 42 \\ \hline \end{array}$

6. $\begin{array}{r} 820 \\ \times\ \ 37 \\ \hline \end{array}$

7. $\begin{array}{r} 16 \\ \times\ 75 \\ \hline \end{array}$

8. $\begin{array}{r} 82 \\ \times\ 45 \\ \hline \end{array}$

9. $\begin{array}{r} 174 \\ \times\ \ 91 \\ \hline \end{array}$

10. $\begin{array}{r} 436 \\ \times\ \ 22 \\ \hline \end{array}$

11. $\begin{array}{r} 738 \\ \times\ \ 63 \\ \hline \end{array}$

12. $\begin{array}{r} 275 \\ \times\ \ 56 \\ \hline \end{array}$

13. $\begin{array}{r} 625 \\ \times\ \ 78 \\ \hline \end{array}$

Estimating Products

You can round the factors to estimate a product. Round each factor to its greatest place value. It is not necessary to round one-digit factors.

Examples

Estimate 67 × 34.

$$\begin{array}{rcr} 34 & \rightarrow & 30 \\ \times\ 67 & \rightarrow & \times\ 70 \\ \hline & & 2{,}100 \end{array}$$

$67 \times 34 \approx 2{,}100$

Estimate 8 × 537.

$$\begin{array}{rcr} 537 & \rightarrow & 500 \\ \times\ 8 & \rightarrow & \times\ 8 \\ \hline & & 4{,}000 \end{array}$$

$8 \times 537 \approx 4{,}000$

PRACTICE

Estimate each product. Show how you find each answer.

1. 88 × 97
2. 162 × 43
3. 372 × 27
4. 18 × 56
5. 6,278 × 32
6. 627 × 12
7. 5,245 × 7
8. 422 × 12
9. 775 × 6
10. 187 × 32
11. 348 × 16
12. 4,862 × 96
13. 65 × 47
14. 143 × 78
15. 97 × 42
16. 588 × 33
17. 4,685 × 8
18. 223 × 12

Reviewing Division

When you divide, you find out how many equal groups there are in a given amount. Division is actually a fast way to subtract equal groups from a total. For example, to find how many 7s there are in 28, we can keep subtracting groups of 7 until there is not enough left to make another group of 7.

$28 - 7 = 21$ $\quad$ $21 - 7 = 14$ $\quad$ $14 - 7 = 7$ $\quad$ $7 - 7 = 0$

We subtracted 7 four times, so there are four 7s in 28.

Here are some important things to keep in mind about division:

- The answer to a division problem is called the **quotient**.
- The number being divided is called the **dividend**.
- The number you are dividing by is called the **divisor**.

dividend ↓ quotient ↓

$56 \div 8 = 7$ $\qquad$ $8\overline{)56}$ = 7

↑ divisor

- A division problem can be written like a fraction. The bar means to divide. Read $\frac{56}{8}$ from top to bottom: 56 divided by 8. $\frac{56}{8} = 7$

- When you divide a number by 1, the quotient is that number.

$756 \div 1 = 1$ $\qquad$ $\frac{34}{1} = 34$ $\qquad$ $1\overline{)56}$ = 56

- If you divide 0 by any number, the quotient will **always** be 0.

$0 \div 8 = 0$ $\qquad$ $\frac{57}{0} = 0$ $\qquad$ $22\overline{)0}$ = 0

- You **cannot** divide by 0. Dividing by zero is undefined.

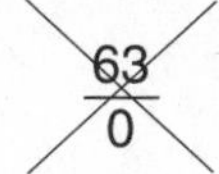

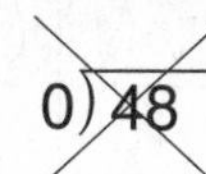

- Multiplication and division are related. Multiplication is the opposite of division. When you multiply, you put equal groups together to find the total. When you divide, you separate the total into equal groups. You can use related multiplication facts to help you with division facts.

7 groups of 9—how many in all? $\quad 7 \times 9 = 63$
63 divided into 7 equal groups—how many in each group? $\quad 63 \div 7 = 9$

9 groups of 7—how many in all? $\quad 9 \times 7 = 63$
63 divided into 9 equal groups—how many in each group? $\quad 63 \div 9 = 7$

- You can use multiplication to check division.
  Multiply the quotient by the divisor to get the dividend.

$\frac{35}{7} = 5$ $\qquad$ $63 \div 9 = 7$ $\qquad$ $8\overline{)48}$ = 6

$7 \times 5 = 35$ $\qquad$ $9 \times 7 = 63$ $\qquad$ $8 \times 6 = 48$

Reviewing Division (continued)

It is important to know basic division facts in order to divide greater numbers with accuracy.

PRACTICE

Find each quotient. Use multiplication to check your answers.

1. $\frac{35}{7} =$ ______
2. $28 \div 4 =$ ______
3. $9\overline{)54}$ ______
4. $81 \div 9 =$ ______
5. $3\overline{)21}$ ______
6. $\frac{18}{3} =$ ______
7. $\frac{45}{5} =$ ______
8. $30 \div 6 =$ ______
9. $8\overline{)56}$ ______
10. $8\overline{)48}$ ______
11. $\frac{24}{6} =$ ______
12. $20 \div 5 =$ ______
13. $15 \div 3 =$ ______
14. $8\overline{)32}$ ______
15. $\frac{25}{5} =$ ______
16. $9\overline{)63}$ ______
17. $\frac{49}{7} =$ ______
18. $24 \div 8 =$ ______
19. $14 \div 7 =$ ______
20. $9\overline{)27}$ ______
21. $\frac{36}{6} =$ ______
22. $\frac{12}{4} =$ ______
23. $16 \div 2 =$ ______
24. $8\overline{)72}$ ______
25. $8\overline{)64}$ ______
26. $\frac{9}{3} =$ ______
27. $16 \div 4 =$ ______
28. $\frac{40}{8} =$ ______
29. $48 \div 6 =$ ______
30. $7\overline{)42}$ ______

Using the Division Frame

A division frame is used to solve division problems. The dividend (number being divided) goes inside the frame. The divisor (number you are dividing by) goes to the left of the frame. The quotient (answer to the division problem) is written above the frame.

- To divide, start on the left side of the dividend with the digit in the greatest place value, and work from left to right. Divide at one digit at a time. Write the quotient directly above the digit being divided.

- After you place the first digit in the quotient, you must write a digit above each remaining digit of the dividend, including the zeros.

4 (hundreds) ÷ 2 = 2 (hundreds)
8 (tens) ÷ 2 = 4 (tens)
0 (ones) ÷ 2 = 0 (ones)

$$2\overline{)480} = 240$$

- If the first digit of the dividend is less than the number you are dividing by, use the first two digits. Write the quotient above the second digit.

42 (tens) ÷ 6 = 7 (tens)
6 (ones) ÷ 6 = 1 (ones)

$$6\overline{)426} = 71$$

4 is less than 6.
Combine 4 with 2 to get 42.

- Sometimes you need to combine digits after you have placed the first number in the quotient. Write a zero in the quotient over the first digit of the combined pair as a placeholder. Remember, after you write the first digit in the quotient, every remaining digit of the dividend must have a digit above it.

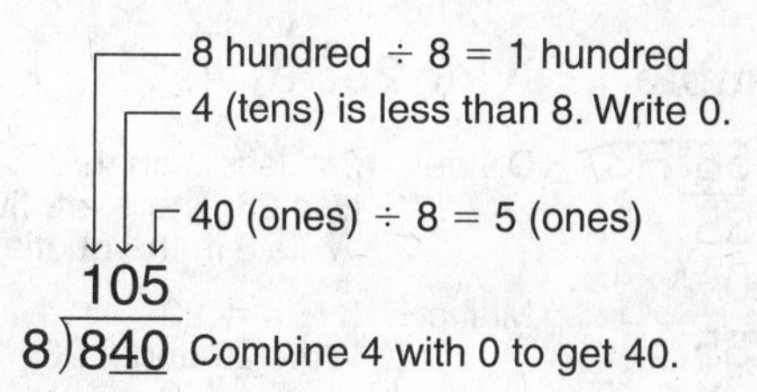

PRACTICE

Find each quotient. Use multiplication to check your answers.

1. $2\overline{)248}$
2. $8\overline{)248}$
3. $3\overline{)369}$
4. $8\overline{)816}$
5. $3\overline{)126}$
6. $3\overline{)324}$
7. $4\overline{)812}$
8. $9\overline{)360}$
9. $6\overline{)366}$
10. $7\overline{)728}$
11. $5\overline{)500}$
12. $3\overline{)327}$
13. What is 628 divided by 2?
14. What is 189 divided by 9?
15. What is 624 divided by 6?
16. What is 48 divided by 4?

Finding a Remainder

One number cannot always be evenly divided by another. For example, if you divide 5 objects into 2 equal groups, there will be 2 groups of 2 and there will be 1 object left over. The amount that is left over, called the **remainder**, is written as part of the quotient. A remainder should always be less than the divisor.

Completing a division problem can take several steps. You divide, then you multiply, and then you subtract. Using a division frame provides an organized way to record each step.

When you finish dividing, check your answer. Multiply the quotient by the divisor, and then add the remainder. The result should match the dividend of the original problem.

Example Divide 37 by 8.

$$\begin{array}{r} 4\text{ R}5 \\ 8\overline{)37} \\ \underline{-32} \leftarrow (4 \times 8) \\ 5 \end{array}$$

Divide: There are four 8s in 37. Write 4 in the quotient.

Multiply: $4 \times 8 = 32$ Write 32 below 37

Subtract: $37 - 32 = 5$

Divide: There is not enough to make another 8. The remainder is 5. Write R 5 in the quotient.

Check

$$\begin{array}{r} 4 \\ \underline{\times 8} \\ 32 \\ \underline{+5} \text{ (remainder)} \\ 37 \end{array}$$

Now look at another example. In this problem, there is one more step: divide, subtract, multiply, then *bring down*. The steps are then repeated.

Example Divide 235 by 4.

$$\begin{array}{r} 58\text{ R}3 \\ 4\overline{)235} \\ \underline{-20} \leftarrow (5 \times 4) \\ 35 \\ \underline{-32} \leftarrow (8 \times 4) \\ 3 \end{array}$$

Divide: 2 is less than 4. Use 23. There are five 4s in 23. Write 5 in the quotient.

Multiply: $5 \times 4 = 20$ Write 20 under 23.

Subtract: $23 - 20 = 3$

Bring down the next digit, 5.

Divide: There are eight 4s in 35. Write 8 in the quotient.

Multiply: $8 \times 4 = 32$ Write 32 under 35.

Subtract: $35 - 32 = 3$

There are no more numbers to bring down. The remainder is 3. Write R 3 in the quotient.

Check

$$\begin{array}{r} {}^{3} \\ 58 \\ \underline{\times 4} \\ 232 \\ \underline{+ 3} \text{ (remainder)} \\ 235 \end{array}$$

PRACTICE

Find each quotient. Show the steps. Be sure to write the remainder as part of the quotient. Then use multiplication to check your answers. Remember to add the remainder.

1.	$2\overline{)27}$	5.	$8\overline{)85}$	9.	$9\overline{)985}$	13.	$6\overline{)840}$
2.	$3\overline{)32}$	6.	$4\overline{)190}$	10.	$8\overline{)1629}$	14.	$3\overline{)2754}$
3.	$7\overline{)79}$	7.	$6\overline{)614}$	11.	$5\overline{)264}$	15.	$4\overline{)865}$
4.	$6\overline{)13}$	8.	$7\overline{)725}$	12.	$7\overline{)1032}$	16.	$9\overline{)7560}$

Estimating Quotients

Estimating a quotient is a quick way to find out *about* what your answer should be. It is also a quick way to check your division. There are different ways to estimate.

- If the first digit, or the first two digits combined, can be divided evenly by divisor, write the quotient in the answer space directly above. Then write zero above each remaining digit of the dividend.

Example Estimate 2,737 ÷ 9. $9\overline{)2{,}737}$ with 300 above

2,737 ÷ 9 ≈ 300

- If you cannot evenly divide the first digit or the first two digits, you can round the dividend. Using the rounded dividend, write the first digit in the quotient and write zero above each remaining digit.

Example Estimate 5,758 ÷ 6.
5,758 can be rounded to 6,000 $6\overline{)5{,}758} \longrightarrow 6\overline{)6{,}000}$ with 1000 above

5,758 ÷ 6 ≈ 1,000

- You can round to a **compatible number**. To find a compatible number, think of multiples of the divisor and round to the multiple closest to the dividend.

Example Estimate 4,137 ÷ 6.
4,137 rounds to 4,000, but 4,000 cannot be evenly divided by 6.
4,137 rounds to 4,100, but 4,100 cannot be evenly divided by 6.
Think of multiples of 6: 6, 12, 18, 24, 30, 36, 42, 48, and so on.

The first two digits of 4,137 are 41, which is close to 42. Round to 4,200. $6\overline{)4{,}137} \longrightarrow 6\overline{)4{,}200}$ with 700 above

4,137 ÷ 6 ≈ 700

PRACTICE

Estimate each quotient. Rewrite the problem to show the numbers you use to make your estimate.

1. $8\overline{)836}$
2. $3\overline{)1{,}852}$
3. $6\overline{)1{,}157}$
4. $5\overline{)3{,}895}$
5. $9\overline{)525}$
6. $4\overline{)2{,}914}$
7. $7\overline{)681}$
8. $8\overline{)148}$
9. $9\overline{)34{,}375}$
10. $7\overline{)46{,}365}$
11. $3\overline{)2{,}268}$
12. $4\overline{)1{,}735}$

Dividing by a Two-Digit Divisor

Before you begin to divide by a two-digit divisor, determine where to place the first digit in the quotient.

- If the first two digits are equal to or greater than the divisor, begin the quotient above the second digit. Once that digit is in place, there will be a digit in the quotient above each remaining digit of the dividend.
- If the first two digits are less than the divisor, use the first three digits. Begin the quotient above the third digit. Each remaining digit of the dividend will need to have a digit above it in the quotient.

Examples

$$15\overline{)315}\quad \text{quotient: xx}$$

$$50\overline{)4{,}260}\quad \text{quotient: xx}$$

The steps for dividing by a two-digit divisor are the same as those for dividing by a one-digit divisor. However, finding the correct digit to place in the quotient each time may require some trial and error. Use your estimating skills. Keep in mind that a product must not be greater than the amount you are dividing. If it is, repeat the division using a lower number for the quotient.

Example

Divide 299 by 23.

```
    13
23)299
  -23← (1 × 23)
    69
  - 69←(3 × 23)
     0
```

Divide: There is one 23 in 29. Write 1 in the quotient.
Multiply: 1 × 23 = 23 Write 23 beneath 29.
Subtract: 29 − 23 = 6
Bring down the next digit.
Divide: There are three 23s in 69. Write 3 in the quotient.
Multiply: 3 × 23 = 69 Write 69 beneath 69.
Subtract: There are no more digits to bring down. There is no remainder.

Check

```
   23
 × 13
   69
 +230
  299
```

299 ÷ 23 = 13

PRACTICE

Find each quotient. Show each step of your division. Use multiplication to check your answers.

1. $20\overline{)260}$ = ______
2. $336 \div 14$ = ______
3. $\frac{800}{16}$ = ______
4. $50\overline{)600}$ = ______
5. $37\overline{)740}$ = ______
6. $688 \div 43$ = ______
7. $\frac{78}{5}$ = ______
8. $20\overline{)484}$ = ______
9. $540 \div 18$ = ______
10. $\frac{1{,}034}{47}$ = ______
11. $450 \div 75$ = ______
12. $24\overline{)720}$ = ______
13. $\frac{85}{17}$ = ______
14. $43\overline{)172}$ = ______
15. $22\overline{)680}$ = ______
16. $1{,}840 \div 80$ = ______
17. $\frac{1{,}900}{38}$ = ______
18. $35\overline{)7{,}035}$ = ______
19. $2{,}100 \div 60$ = ______
20. $\frac{104}{13}$ = ______
21. $19\overline{)7{,}638}$ = ______
22. $880 \div 40$ = ______
23. $14\overline{)5{,}628}$ = ______
24. $50\overline{)300}$ = ______
25. $\frac{1{,}050}{35}$ = ______
26. $260 \div 65$ = ______
27. $15\overline{)480}$ = ______
28. $\frac{589}{14}$ = ______
29. $\frac{57}{8}$ = ______
30. $912 \div 6$ = ______

Computation with Whole Numbers Skills Checkup

Circle the letter of the answer to each problem. Use your estimation skills to eliminate unreasonable answers before you begin to calculate the answer.

1. 315×9

   A 2,715
   B 2,735
   C 2,845
   D 2,835
   E None of these

2. $272 \div 8 =$

   F 34
   G 30 R2
   H 44
   J 32
   K None of these

3. $231 \times 50 =$

   A 1,155
   B 11,550
   C 10,550
   D 1,055
   E None of these

4. $3{,}801 \times 13$

   F 39,253
   G 15,204
   H 47,413
   J 49,413
   K None of these

5. $14\overline{)420}$

   A 3
   B 33
   C 29
   D 40
   E None of these

6. $3\overline{)672}$

   F 224
   G 220 R2
   H 230
   J 223
   K None of these

7. $\$14.60 - \$7.85$

   A $ 7.25
   B $ 6.75
   C $22.45
   D $ 6.85
   E None of these

8. $35 \times 8 =$

   F 280
   G 240
   H 28
   J 285
   K None of these

9. $842 \div 6 =$

   A 14 R2
   B 107
   C 140 R2
   D 130 R2
   E None of these

10. $3\overline{)105}$

    F 35
    G 30 R5
    H 3 R5
    J 33
    K None of these

11. $\begin{array}{r} \$8.95 \\ +\ \$5.25 \\ \hline \end{array}$

A $14.10
B $14.15
C $13.10
D $13.20
E None of these

12. $435 \times 60 =$

F 2,610
G 25,800
H 42,600
J 26,100
K None of these

13. $84 \div 12 =$

A 70
B 7
C 60
D 7 R2
E None of these

14. $\begin{array}{r} 53 \\ \times\ 21 \\ \hline \end{array}$

F 1,013
G 159
H 1,113
J 213
K None of these

15. $15\overline{)480}$

A 32
B 320
C 25
D 250
E None of these

Circle the best estimate.

16. 27×32

A 9 C 90
B 900 D 9,000
E None of these

17. $576 \div 8$

F 7 H 70
G 700 J 7,000
K None of these

18. 735×45

A 350 C 3,500
B 35,000 D 350,000
E None of these

19. $2{,}753 \div 14$

F 600 H 200
G 400 J 100
K None of these

20. $4{,}827 - 1{,}095$

A 1,000 C 2,000
B 3,000 D 4,000
E None of these

Problem Solving

Following a Five-Step Plan

Adding, subtracting, multiplying, and dividing are only one part of the problem-solving process. When solving word problems, follow the five steps below.

Identify the question.
- Read the problem. Make sure you understand all of the words and ideas.
- Ask yourself what you have to find out to solve the problem.
- Try to retell the problem in your own words.

Identify the information to use.
- Read the problem again to see what information has been given.
- Eliminate information you do not need.
- Decide if there is information missing. If so, is it information you can get?
- Look for signal words that can help you decide what operation(s) to use.

Make a plan.
- Think about what you need to do first.
- Decide what operation or operations you will use.
- Write a number sentence.

Carry out your plan.
- Do the computation. Be sure to record all your work.
- Label the answer if a label is needed.

Check your results.
- Check your computation.
- Make sure you answered the question asked in the problem.
- Check to see that your answer is reasonable. Does it make sense?

Here are additional things that can help you with word problems.

- Read the problem through more than one time before you begin to solve.
- Underline the question being asked.
- Cross out any information that is not needed.
- List the information in the problem that can help you find the answer to the question.

The chart below contains signal words and phrases that can help you decide which operation to use. Remember, phrases such as *about how much* or *approximately how much* generally signal that an estimate is called for.

Signal Words

Addition	Subtraction	Multiplication	Division
plus sum total added to all together in all combined increased by	minus difference take away subtract left over how much change more than less than decreased by	times multiplied by product twice, three times, and so on apiece each	divide apiece each per quotient divided by how many equal groups shared equally separated into equal groups

Identifying the Question to Answer

Before you start to solve a problem, make sure you know what you are trying to find. Many people find it helpful to underline the question.

PRACTICE

Circle the letter of the question that best describes what each problem asks you to find.

1. Last year Mr. Jay traveled a total of 457,975 miles for business. The year before that, it was 260,783 miles. How many more miles did Mr. Jay travel last year than the year before last?

   A What was the difference in the number of miles Mr. Jay traveled in the two years?
   B In which year did Mr. Jay travel a greater number of miles?
   C What was the total number of miles traveled in the two years?
   D None of these

2. Lucy has already lost 25 pounds on her diet. Her friend Penny has lost 20 pounds. Lucy wants to lose a total of 45 pounds. How many more pounds does she have to lose?

   F Who has lost more weight?
   G In all, how many pounds have Lucy and Penny lost?
   H How much weight does Lucy still have to lose?
   J None of these

3. Three friends went to dinner. The bill was $75 plus $6.38 for tax. If they added a tip and then shared the cost equally, what was each person's share?

   A What was the total cost of the meal?
   B How much should they have left for a tip?
   C How much did each person pay?
   D None of these

4. Bill's son has math and reading homework every night. He spends 20 minutes on the reading and then another 30 minutes doing math. How many minutes each week does he spend on his reading assignments?

   F How much time does Bill's son spend on his reading assignments each week?
   G Does Bill's son spend more time on reading or math?
   H What is the total amount of time Bill's son spends doing his assignments each night?
   J None of these

5. Some friends are planning a wedding shower for Carrie. They have invited 25 people to the shower. So far, 12 people have called to say they would attend, and 4 have said they would not be able to come. How many people have not responded yet?

   A How many people will be at the shower?
   B How many more people said they would attend than would not?
   C How many people still have to let the friends know whether they will attend?
   D None of these

Identifying Too Much or Too Little Information

It is important to read through a problem very carefully before you start working to solve it. There might not be enough information for you to be able to solve it, or there might be information that you do not need.

PRACTICE

Problems 1–4 contain unnecessary information. Draw a line through the information that is not needed to solve the problem.

1. Becky agreed to make 12 dozen cookies for the sale. Her sister ended up making 4 dozen. How long will it take Becky to mix 6 batches of cookies if it takes her 10 minutes to mix 1 batch?

2. In the past two years, 250 people have moved to Carville. Of 3,000 people working there, 156 are employed at the paint factory. How many workers in Carville do not work at the paint factory?

3. Rachel wants to take the 12 girls in her scout troop to Washington, D.C. The airfare would be $165 per scout. The hotel would cost $80 per person per night. How much would the airfare for the girls in the troop cost?

4. Ron is 65 years old. His wife is 3 years younger than he is. When she was 25, she had their first child. How old was Ron's wife on that child's 10th birthday?

Problems 5–8 do not give enough information to be solved. Tell what additional information is needed.

5. Phu rented a car for 3 days. How much did he pay for the rental?

____

____

6. Jill's company treated the secretaries in the office to lunch. What was the total cost of the lunches?

____

____

7. Cameron drove to his mother's house for Thanksgiving dinner. How long did the drive take?

____

____

8. If Brandy chooses the carpet that sells for $25 per square yard, how much will it cost her to carpet her bedroom?

____

____

Choosing an Operation and Solving the Problem

You know the question to answer, and you have the information needed to solve the problem. Now, determine what operation to use. Look for signal words that can help you. (See page 32.) Then you can write a number sentence to show your plan.

PRACTICE

Circle the letter for the operation you would use to solve each problem. Then write a number sentence and solve the problem. The first one is done for you.

1. A plumber worked for 3 hours at Wayne's house. He charged $180 for the job. How much did the plumber charge per hour?

   A add
   B subtract
   C multiply
   (D) divide

   $180 ÷ 3 = $60

2. The gas tax in our town is 23 cents per gallon. If you buy 20 gallons of gas, how much tax do you pay?

   F add
   G subtract
   H multiply
   J divide

   ____

3. Lance loaded 89 boxes into his delivery truck in the morning. At 2 o'clock that afternoon he had 36 boxes left on the truck. How many boxes had he delivered?

   A add
   B subtract
   C multiply
   D divide

   ____

4. Ed's Remodelers charged $7,000 to remodel a bathroom. After he bought materials and paid his workers, Ed had a profit of $3,259. How much were his costs for this job?

   F add
   G subtract
   H multiply
   J divide

   ____

5. Herman bought a sweater for $48.58. There was a tax of $3.60 on the sweater. What was the total cost of the sweater?

   A add
   B subtract
   C multiply
   D divide

   ____

6. Four friends took Marci out to dinner for her birthday. The bill came to $94. If the four shared the cost equally, how much was each person's share?

   F add
   G subtract
   H multiply
   J divide

   ____

Solving Two-Step Word Problems

Some problems require two steps to be solved. For example, you might have to add two numbers, then multiply the sum by a number. Any combination of operations is possible. There is no special trick to solving two-step problems—just read them carefully, think about what is happening in the problem, and look for signal words.

PRACTICE

Each problem below needs two steps to be solved. Circle the letter of the choice that describes how to solve the problem.

1. Mrs. Carr's class is taking a field trip and 5 parents have offered to help drive. Four parents have vans, and each can take 9 children. The fifth parent has a car that can carry only 4 children. Together, how many children can these parents drive?

   A Add 9 and 4. Then multiply by 5.
   B Multiply 9 by 4. Then add 4.
   C Add 4 and 9. Then add 4.

2. Terry has $6.84 in his wallet. He gets $40 from an ATM. Then he spends $15.95. How much money does he have left?

   F Add $6.84 and $40. Then subtract $15.95.
   G Subtract $6.84 from $40. Then subtract $15.95.
   H Add $6.84 and $15.95. Then subtract the sum from $40.

3. Wendell buys 3 pounds of hamburger for $2.56 a pound, plus a box of cereal for $3.95. How much does he spend? *(Ignore the tax.)*

   A Add $3.95 and $2.56. Then multiply by 3.
   B Add 3 and $2.56. Then add $3.95.
   C Multiply $2.56 by 3. Then add $3.95.

Solve each two-step problem below.

4. Ellen knows that it will take her 15 minutes to mix a batch of cookies, and then 10 minutes per pan to bake them. If a batch of cookies makes 3 pans, how long will it take Ellen to mix and bake them all?

5. Eric buys 3 pounds of chocolates for $4.50 a pound. The tax on his purchase is $1.25. How much does he spend?

6. Tina had $2,546 in a savings account. She deposited a $520 dollar check, and then took out $40 in cash. What was the balance in her savings account?

7. Jorge has 28 miniature cars and 16 stuffed animals that he has saved from his childhood. He divides the toys evenly among his 4 children. How many toys does each child get?

Using Estimation to Solve Word Problems

The words *about, approximately,* and *almost* usually signal that a problem calls for estimation. Set up the problem and then round the numbers *before* you do any calculating.

PRACTICE

Circle the letter of the number sentence that shows the *best* way to solve each problem.

1. In an election, 7,845 people voted for Miller, and 3,182 voted for her opponent. About how many people voted in that race?

   A 7,845 + 3,182 = ______
   B 7,000 + 3,000 = ______
   C 8,000 + 3,000 = ______

2. Each morning, Greg's train ride takes 27 minutes. It takes about 12 minutes for him to get to the train, and another 7 minutes to get from the train to his office. Approximately how much time does Greg spend traveling to work?

   F 27 + 12 + 7 = ______
   G 20 + 10 + 10 = ______
   H 30 + 10 + 10 = ______

3. Glenn earns $32,650 a year, plus about $3,000 a year in bonuses. About how much does he earn per month?

   A ($32,650 + $3,000) ÷ 12 = ______
   B ($33,000 + $3,000) ÷ 12 = ______
   C ($32,000 + $3,000) ÷ 12 = ______

4. Naomi spent 12 minutes on a stair machine, 9 minutes on a bike, and 30 minutes doing aerobics. About how long did she exercise?

   F 15 + 10 + 30 = ______
   G 15 + 10 + 50 = ______
   H 10 + 10 + 30 = ______

Use estimation to solve each of the following problems.

5. It took Linda 3 hours to drive 178 miles. Approximately how many miles did she drive per hour? *(Round 178 to the tens place.)*

6. This month, Carlyle spent $72.50 to get his car tuned up, $125.12 to get new brake pads, and $119.95 to get a new alternator. To the nearest dollar, how much did he spend?

7. Arlene took 6 dresses to the dry cleaner. The cleaner charges $2.95 per dress. To the nearest dollar, how much will Arlene be charged?

8. George has a library book that is 23 days overdue. Library fines are 25 cents a day. About how much will George owe? *(Round each number to the tens place.)*

Problem-Solving Skills Checkup

Study the mileage chart shown below. Then do Numbers 1–3.

Mileage Chart	
Kirkwood to St. Louis	12 miles
St. Louis to Fort Leonard Wood	128 miles
St. Louis to Rolla	106 miles
St. Louis to Dixon	128 miles
Dixon to Fort Leonard Wood	20 miles
Dixon to St. Robert	17 miles
St. Robert to Fort Leonard Wood	5 miles

1. How many miles is the trip from St. Louis to Fort Leonard Wood if you stop off in Dixon on the way?

   A 128 miles
   B 148 miles
   C 108 miles
   D Not enough information is given.

2. Kirkwood lies between St. Louis and Rolla on Highway 44. How far is Kirkwood from Rolla?

   F 118 miles
   G 94 miles
   H 128 miles
   J This cannot be determined.

3. You are traveling to Fort Leonard Wood from Dixon. How many miles would you add to your trip if you stopped off in St. Robert along the way?

   A 0 miles
   B 3 miles
   C 2 miles
   D 5 miles

4. Lane must buy 3 hot dogs that cost $1.99 each. Which of these number sentences could he use to find the total cost?

   A $\$1.99 + 3 =$
   B $\$1.99 \times 3 =$
   C $\$1.99 \div 3 =$
   D $\$1.99 - 3 =$

5. Luke and Garrett earn $500 moving a family across town. They use $40 of that money to pay for renting the moving van. If they split the rest of the money evenly, how much money will each man get?

   F $250
   G $210
   H $460
   J $230

6. Irena sold 25 makeup kits, and she got $10 for each one. She spent $100 to buy the samples that she shows to her customers. What must Irena do to figure out how much money she has made?

   A Multiply $100 by 25. Then subtract $10 from the product.
   B Divide $100 by $10. Then multiply the quotient by 25.
   C Multiply $10 by 25. Then subtract $100 from the product.
   D Multiply $10 by 25. Then add $100 to the product.

7. Tyra earns $2,150 every month. She pays $450 each month for rent and makes sure she puts $25 into her savings account. How much does Tyra have left to cover her other expenses for the month?

   F $1,750 H $1,775
   G $1,675 J $1,725

8. Vera sold a set of dishes through the Internet Sales Company for $85. This was $43.85 less than she had paid for the dishes. How much did Vera pay for the dishes?

   A $128.85 C $ 41.15
   B $ 43.00 D $ 75.00

9. Alex bought 3 cases of cola, 1 case of ginger ale, and 1 case of fruit-flavored soda for his annual barbecue. There were 24 cans in each case. What was the total number of cans of soda Alex bought?

   F 72 H 100
   G 96 J 120

10. Sidney paid $87.63 for food for her party. She spent another $36.88 for beverages. She also bought paper plates, plastic cups, and napkins, which cost $32.75. What was the total cost of these party items?

    A $157.26 C $120.38
    B $124.51 D $159.26

11. The All Welcome Inn on Highway 5 first opened for business in 1857. The inn has remained in business and has been run by members of the same family ever since. In 2007, the family is planning to celebrate a special anniversary for the inn. How many years will it have been in operation?

    F 143 H 140
    G 150 J 250

12. If the average adult gets 7 hours of sleep a night, how many hours of sleep does he or she get in one year? (1 year = 365 days)

    A 2,245 C 2,735
    B 2,555 D 2,475

13. Theo figures that he spends about $4,056 each year on insurance, maintenance, and gasoline for his car. He spends time each weekend cleaning and polishing the car. Approximately how much are Theo's car expenses each month?

    F $200 H $600
    G $400 J $800

14. The local theater is remodeling, and the seats are being replaced. If there are 36 rows with 25 seats in each row, how many new seats will be needed?

    A 870 C 890
    B 880 D 900

Decimals

Reviewing Decimal Place Value

In our decimal place-value system, we write amounts *less than 1* as **decimal fractions**, or more simply, as **decimals**. We use a **decimal point** to separate decimals from whole numbers, and just as in whole numbers, each digit's value is determined by its place. Notice that the names of the place-value columns on either side of the ones column are similar, but that the decimal columns end with *–ths*.

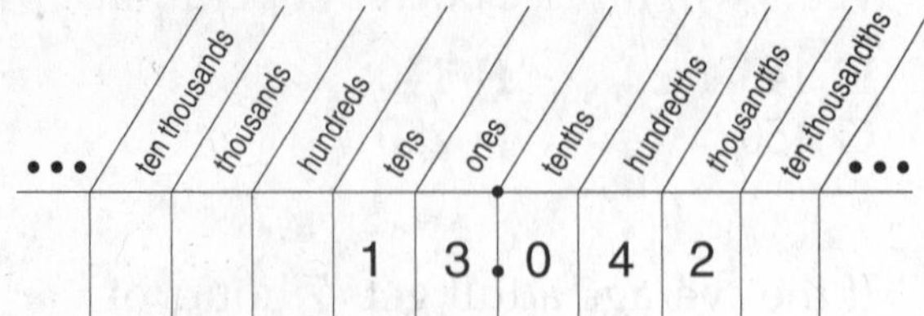

The number 13.042 is greater than 13, but it is less than 14. In 13.042, the 0 represents zero tenths, the 4 represents four hundredths, and the 2 represents two thousandths.

To read or write a decimal in words, say the digits to the right of the decimal point as if they were a whole number and then add the name of the place-value column of the digit farthest to the right.

- The decimal portion of 13.042 is *forty-two thousandths.*
- The entire amount is *thirteen **and** forty-two thousandths.* The word *and* represents the decimal point.

PRACTICE

Write the place value of the column for the digit named.

1. the 3 in 1.93 ____________
2. the 4 in 5.432 ____________
3. the 9 in 0.09 ____________
4. the 6 in 4.026 ____________

Write the number.

5. six and three thousandths ______
6. one hundred thousandths ______
7. two and three hundredths ______
8. twelve thousandths ______

Write the number in words.

9. 2.04 ____________________________
10. 0.140 ____________________________
11. 62.02 ____________________________
12. 15.207 ____________________________
13. 9.09 ____________________________

Comparing Decimals

The value each place-value column represents decreases as you move to the right. One hundredth is less than 1 tenth, and 1 thousandth is less than 1 hundredth, and so on.

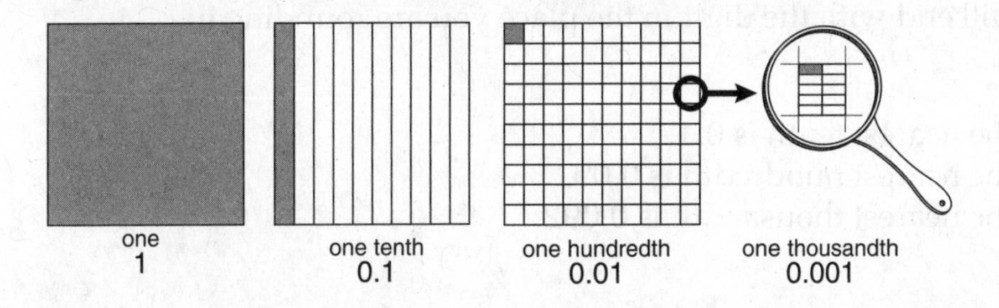

To compare two decimal numbers, start by lining up the decimal points. Then, moving from left to right, compare the digits in each column.

Examples

0.02
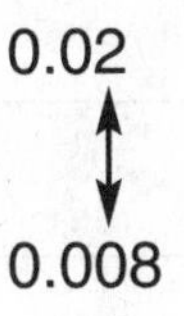
0.008

2 hundredths is greater than 0 hundredths.
$0.02 > 0.008$

2.105
↕
12.182

1 ten is greater than no tens.
$12.182 > 2.105$

15.063
↕
15.01

6 hundredths is greater than 1 hundredth.
$15.063 > 15.01$

PRACTICE

Circle the greater number in each pair.

1.	0.1	0.0283
2.	0.0089	0.13
3.	1.123	0.059
4.	12.32	12.50

Circle the least number in each set.

5.	0.02	0.005	0.1
6.	1.013	1.301	1.32
7.	0.98	0.99	1.0
8.	4.03	14.001	4.014

Write the numbers in each set in order from least to greatest.

9.	34.487	34.478	38.744	37.448
10.	9.909	6.873	12	10.50
11.	8.3	8.03	8.35	8.035
12.	0.865	0.79	0.9	0.897

Write the numbers in each set in order from greatest to least.

13.	4.051	4.020	4.20	4.015
14.	0.623	0.20	0.63	0.9

Rounding Decimals

Rounding decimal numbers is similar to rounding whole numbers. To round to a particular place value, look at the digit to the *right* of that place.

- If the digit to the right is 5 or greater, the digit in the place you are rounding to will increase by one.
- If the digit to the right is 4 or less, the place you are rounding to will remain unchanged.

The decimal number will end with the digit in the place you are rounding to.

Examples

0.0916 rounded to the nearest tenth is 0.1.
0.0916 rounded to the nearest hundredth is 0.09.
0.0916 rounded to the nearest thousandth is 0.092.

PRACTICE

Round each number to the nearest tenth.

1. 0.842 ____________
2. 3.9356 ____________
3. 7.017 ____________
4. 0.96 ____________
5. 15.96 ____________
6. 9.96 ____________

Round each number to the nearest hundredth.

7. 0.3749 ____________
8. 2.995 ____________
9. 14.003 ____________
10. 12.483 ____________
11. 6.781 ____________
12. 10.096 ____________

Round each number to the nearest thousandth.

13. 2.0045 ____________
14. 7.49582 ____________
15. 0.01234 ____________
16. 15.8275 ____________
17. 0.9895 ____________
18. 0.9999 ____________

Circle the number in each set that shows the first number rounded to the nearest hundredth.

19. 4.375 4.37 4.36 4.38
20. 0.997 0.98 0.99 1.00
21. 0.993 0.98 0.99 1.00

Adding Decimals

To add decimals, add tenths to tenths, hundredths to hundredths, and so on. Write addends one below another and line up the decimal points to line up the digits in columns. Then start at the right and work to the left. Add one column at a time, regrouping as needed. Be sure to write a decimal point in the answer.

Example Add 41.3 + 2 + 0.093.

```
 41.3   ← Line up the decimal points.
  2.
+ 0.093
-------
```

```
 41.300 ← You can write zeros to fill
  2.000 ← the empty places in each column.
+ 0.093
-------
 43.393
```

41.3 + 2 + 0.093 = 43.393

PRACTICE

Add. First write each problem in column form with decimal points lined up. Check your answers using rounded values.

1. 32 + 0.01 = ____________
2. 0.06 + 0.17 = ____________
3. 1.5 + 0.023 = ____________
4. 1.03 + 0.56 = ____________
5. 0.09 + 0.015 = ____________
6. 0.035 + 0.29 + 1.1 = ____________
7. 7.15 + 0.25 = ____________
8. 9.08 + 0.15 = ____________
9. 15.129 + 8.08 = ____________
10. 0.031 + 10 + 2.2 = ____________

Subtracting Decimals

To subtract decimals, subtract tenths from tenths, hundredths from hundredths, and so on. Write the number being subtracted below the starting amount with decimal points and digits lined up in columns. Write a zero in each empty column to make the number of digits in each number equal. Then start at the right and work to the left and subtract. Regroup as needed. Be sure to write a decimal point in the answer.

Example Subtract 12.7 − 0.16.

```
          ┌── Line up the          6 10
  12.7       decimal points.    12.7̸0̸ ←── Write a zero in any empty column.
− 0.16                        −  0.16
──────                        ───────
                                12.54
```

12.7 − 0.16 = 12.54

To subtract from a row of zeros, a shortcut is to write a digit one less than the first nonzero digit from the left. Next, write a ten over the zero farthest to the right and write nines over the other zeros. Then you are ready to subtract.

Example

```
                      3  9 9 9 10
  24.0000            24̸.0̸0̸0̸0̸
−  0.0028          −  0.0028
─────────          ─────────
                     23.9972
```

PRACTICE

Subtract. Then check your answers using addition.

1. 3.1 − 0.1 = ______
2. 0.765 − 0.003 = ______
3. 1.5 − 0.015 = ______
4. 0.5 − 0.07 = ______
5. 4.25 − 1.1 = ______
6. 15.6 − 0.025 = ______
7. What is \$35.19 − \$4.07?
8. What is 47 cents subtracted from \$1.15?
9. What is 0.03 subtracted from 0.264?
10. What is 0.8 subtracted from 9?
11. What is 3 cents subtracted from \$15.61?
12. What is 0.017 subtracted from 5?

Multiplying Decimals

When you multiply decimals, do *not* line up the decimal points. Count the digits to the right of the decimal point in each factor—that total determines how many digits should be to the right of the decimal point in the answer. Then multiply as if you were multiplying whole numbers. When you finish multiplying, place the decimal point in the product.

Example

Digits to right of decimal point

```
  2.24 ←   2 digits          1 2
×    5 ← +0 digits          2.24
           2 digits       ×    5
                           11.20
```

To place the decimal point in the answer, start at the right.
Count 2 digits as you move to the left.

You can often use estimation to check your answer and make sure it is reasonable. Since 2.24 is between 2 and 3, it makes sense that 5 × 2.24 is between 5 × 2 and 5 × 3, or between 10 and 15. The answer of 11.20 is reasonable.

PRACTICE

Complete Numbers 1–10 by writing the decimal point in the product. Use estimation to check your answers.

```
1.   0.14        7.    0.4
    ×   2            ×   3
       28               12

2.    0.7        8.   15.6
    × 0.3            × 0.2
       21              312

3.    1.5        9.   0.35
    ×   5           × 0.75
       75              175
                      2450
                      2625

4.   2.01       10.   63.8
    × 0.3            × 1.2
      603             1276
                      6350
5.   21.2             7626
    × 0.4
      848

6.   0.92
   × 0.12
      184
      920
     1104
```

Find each product.
Remember that in math, the word *of* means *multiply*.

11. 7 × 0.16 = ______

12. 12 × 1.2 = ______

13. 0.5 × 4.6 = ______

14. 8.45 × 160 = ______

15. What is 0.02 of 20?

16. What is 0.14 of 100?

17. What is 5 tenths of $2.50?

18. What is 3 tenths of 300?

Multiplying Decimals (continued)

Sometimes there are not enough digits in the product to place the decimal point. When that happens, write zeros at the **left** of the product to create the number of places needed. Then write the decimal point. Look at the following examples.

Examples

$$\begin{array}{rl} 0.3 & \leftarrow \text{ 1 place} \\ \times\ 0.2 & \leftarrow +1 \text{ place} \\ \hline & \text{2 places} \end{array} \qquad \begin{array}{r} 0.3 \\ \times\ 0.2 \\ \hline 0.06 \end{array} \qquad \begin{array}{rl} 0.15 & \leftarrow \text{ 2 places} \\ \times\ 0.03 & \leftarrow +2 \text{ places} \\ \hline & \text{4 places} \end{array} \qquad \begin{array}{r} 0.15 \\ \times\ 0.03 \\ \hline 0.0045 \end{array}$$

PRACTICE

Multiply. Then check your work using estimation.

19. $0.4 \times 0.2 =$ ______

20. $0.03 \times 0.4 =$ ______

21. $0.19 \times 0.2 =$ ______

22. $\begin{array}{r} 0.025 \\ \times\ 0.05 \\ \hline \end{array}$

23. $\begin{array}{r} 0.62 \\ \times\ 0.003 \\ \hline \end{array}$

24. $\begin{array}{r} 0.04 \\ \times\ 0.012 \\ \hline \end{array}$

25. What is 0.7 of 0.005?

26. What is 0.3 of 1.2 ounces?

27. What is 2 hundredths of 12.06?

28. What is 3 hundredths of 0.2?

29. What is one tenth of 0.5?

30. What is 3 tenths of 0.09?

31. What is 25 hundredths of 40?

32. What is one thousandth of 6?

Dividing a Decimal by a Whole Number

To divide a decimal by a whole number, first write a decimal point in the quotient directly above the decimal point in the dividend. Then divide as you do with whole numbers.

Example Divide 6.04 by 2.

$$2\overline{)6.04}$$

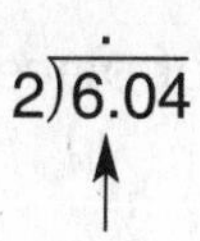

Write the decimal point in the quotient.

$$\begin{array}{r} 3.02 \\ 2\overline{)6.04} \end{array}$$

Then divide as usual.

$6.04 \div 2 = 3.02$

Every digit to the right of the decimal point in the dividend should have a digit above it in the quotient. Write zeros above digits in the quotient as placeholders when dividing numbers that are less than the divisor.

$$\begin{array}{r} 2.103 \\ 3\overline{)6.309} \end{array} \qquad \begin{array}{r} 0.004 \\ 5\overline{)0.020} \end{array}$$

PRACTICE

Find each quotient. Check your answers using multiplication.

1. $2\overline{)4.08}$
2. $3\overline{)0.189}$
3. $5\overline{)80.5}$
4. $6\overline{)\$12.24}$
5. $7\overline{)210.63}$
6. $32\overline{)3.360}$
7. $10.5 \div 21 =$ ______
8. $9.207 \div 9 =$ ______
9. $37.1 \div 7 =$ ______
10. $0.912 \div 3 =$ ______
11. $5.040 \div 40 =$ ______
12. What is 1.35 divided by 5?
13. What is 0.1053 divided by 13?
14. What is $42.84 divided among 6 people?
15. What is 0.46 divided by 0.02?

Dividing by a Decimal

When a problem calls for dividing by a decimal, you must adjust the problem to make the divisor a whole number.

- Move the decimal point of the divisor to the right as many places as needed to make the divisor a whole number.
- Move the decimal point of the dividend to the right the same number of places.*
- Write the decimal point in the new position in the quotient.
- Divide as usual.

Example Divide 0.528 by 0.33.

```
           .                      1.6              Check
0.33)0.528   ——→         .33)0.528                0.33
                              - 33               × 1.6
                               198                 198
                              -198               + 330
                                 0               0.528
```

$0.528 \div 0.33 = 1.6$

To check your answer, multiply the quotient by the **original** divisor. The product should be the **original dividend**.

*Note: "Moving the decimal point to the right" is the same as multiplying by a power of ten. Moving the decimal point one place to the right is equivalent to multiplying by 10, moving it two places is equivalent to multiplying by 100, and so on. To keep the same relationship between the dividend and the divisor, you must multiply, or move the decimal point, in both the divisor and the dividend.

PRACTICE

Find each quotient. Use multiplication to check your answers.

1. 0.8)76.80
2. 0.07)0.42
3. 0.32)2.88
4. 0.3)14.88
5. 1.6)7.68
6. 0.08)1.92
7. 0.12)0.276
8. 0.20)5.26
9. What is 53.9 ÷ 0.07?
10. What is 42.84 divided by 0.06?
11. What is 35.75 ÷ 6.5?
12. What is 0.488 ÷ 0.4?

If the number you are dividing is a whole number, you will need to write a decimal point and zeros.

- Place a decimal point to the right of the ones digit in the dividend.
- Write zeros to the right of the decimal point.
- Move the decimal point to the right as many places as needed.
- Divide.

Example Divide 28 by 0.4.

$$0.4\overline{)28} \rightarrow 0.4\overline{)28.0} \rightarrow .4\overline{)28.0}$$

$$\begin{array}{r} 70. \\ .4\overline{)28.0} \\ -28 \\ \hline 00 \end{array}$$

Check

$$\begin{array}{r} 70 \\ \times 0.4 \\ \hline 28.0 \end{array}$$

$28 \div 0.4 = 70$

PRACTICE

Find each quotient. Check your answers using multiplication.

13. $0.9\overline{)360}$
14. $0.04\overline{)13}$
15. $1.6\overline{)96}$
16. $0.012\overline{)144}$
17. $0.03\overline{)189}$
18. $81 \div 0.03 =$ ______
19. $768 \div 2.4 =$ ______
20. $574 \div 0.2 =$ ______
21. $168 \div 1.4 =$ ______
22. $375 \div 1.25 =$ ______

Using Decimals to Solve Word Problems

Circle the letter of the number sentence or expression that shows how to solve the problem. Then solve it.

1. A 4-ounce bottle of perfume costs \$26.56. How much does 1 ounce of the perfume cost?

   A \$26.56 + 4 = ______
   B \$26.56 − 4 = ______
   C \$26.56 × 4 = ______
   D \$26.56 ÷ 4 = ______

2. Zora bought 8 bottles of detergent. Each contains 14.3 ounces. How many ounces did she buy in all?

   F 14.3 + 8 = ______
   G 14.3 − 8 = ______
   H 14.3 × 8 = ______
   J 14.3 ÷ 8 = ______

3. An uncooked roast weighed 5.3 pounds. After it was cooked, it weighed 3.7 pounds. How much did the roast shrink while cooking?

   A 5.3 + 3.7 = ______
   B 5.3 − 3.7 = ______
   C 5.3 × 3.7 = ______
   D 5.3 ÷ 3.7 = ______

4. A book costs \$14.15. Sales tax is 9 cents per dollar. What is the total cost of the book?

   F (\$14.15 × 0.09) + \$14.15
   G \$14.15 + \$0.09
   H \$14.15 × \$0.09
   J (\$14.15 ÷ \$0.09) + \$14.15

Solve each word problem below. Use estimation to check your answers.

5. When Dan began his road trip, the odometer on his car read 1,256.7 miles. When he finished the trip, it read 1,814.9 miles. How far did Dan drive?

6. Renee finished a 7.8-mile hike in 2 hours. If she walked at a steady rate, how far did she walk each hour?

7. French fries cost \$0.75 per order. Hamburgers cost \$3.59 each. How much would a hamburger and 3 orders of fries cost?

8. A 12-ounce bag of potato chips costs \$3.72. A 10-ounce bag of barbecue chips costs \$2.90. How much more do the potato chips cost per ounce than the barbecue chips? (*Hint:* This is a two-step problem.)

9. With your telephone plan, your long distance calls are \$0.12 a minute. If you have a charge of \$2.88 for a call, how many minutes did you talk?

10. Zora bought 6 towels for \$6.59 each on sale. The tax came to \$3.16. What was the total cost of the towels?

Decimals Skills Checkup

Circle the letter for the correct solution to each problem. Try crossing out unreasonable answers before you start to work on each problem.

1. $9.15 + 0.058 =$

   A 9.73 C 9.108
   B 9.208 D 0.973
   E None of these

2. $0.103 - 0.07 =$

   F 0.133 H 0.033
   G 0.005 J 0.05
   K None of these

3. $15.42 \times 0.003 =$

   A 46.26 C 4.626
   B 0.4626 D 0.04626
   E None of these

4. $2.1\overline{)462}$

   F 220 H 2.2
   G 0.022 J 22
   K None of these

5. $2.79 \div 31 =$

   A 9 C 0.9
   B 0.09 D 90
   E None of these

6. $95 - 0.015 =$

   F 80 H 95.085
   G 94.085 J 94.095
   K None of these

Use the following information to do Numbers 7–10.

At the end of the eighth grade, Lin was 3.6 feet tall. Over the next 2 years he grew 1.2 feet.

7. How tall was Lin at the end of the tenth grade?

   A 4.4 feet
   B 4.8 feet
   C 2.4 feet
   D 4.32 feet

8. On average, how much did Lin grow per month during the 2-year period?

   F 1 foot
   G 0.05 feet
   H 0.5 feet
   J 1.2 feet

9. One meter is about 3.3 feet. To find Lin's height in meters at the end of the eighth grade, which should you do?

   A Add 3.6 and 3.3.
   B Multiply 3.6 by 3.3.
   C Subtract 3.3 from 3.6.
   D Divide 3.6 by 3.3.

10. In eighth grade, the tallest boy in Lin's class was 6.2 feet tall. What was the difference between his height and Lin's height?

    F 2.6 feet
    G 3.6 feet
    H 3.4 feet
    J 5.0 feet

Decimals Skills Checkup (continued)

11. $\begin{array}{r} 9.82 \\ +\ 0.3 \\ \hline \end{array}$

A 9.83
B 10.12
C 11.12
D 12.82
E None of these

12. $4.35 × 11 =

F $47.85
G $8.70
H $87.00
J $478.50
K None of these

13. 2.04 − 0.009 =

A 2.031
B 2.31
C 2.0391
D 2.0301
E None of these

14. $\begin{array}{r} \$15.80 \\ -\ 9.65 \\ \hline \end{array}$

F $7.15
G $5.15
H $6.25
J $6.15
K None of these

15. 0.95 + 0.79 =

A 1.64
B 1.74
C 0.0174
D 0.164
E None of these

16. Which of these number sentences could be used to find one-tenth of 46?

F 10 × 46 =
G 46 ÷ 0.1 =
H 46 × 0.01 =
J 46 × 0.1 =

17. What is 6.075 in words?

A six hundred and seventy-five thousandths
B six and seventy-five hundredths
C six and seventy-five thousands
D six and seventy-five thousandths

18. Which set of decimal numbers is in order from least to greatest?

F 0.19 0.85 1.003 0.091
G 0.091 0.19 0.85 1.003
H 1.003 0.19 0.85 0.091
J 0.19 1.003 0.85 0.091

19. A decimal number has 0 ones, and then uses each of the digits 9, 3, 0, and 1 exactly once. What is the least possible value for the decimal number?

A 0.0931
B 0.0139
C 0.1390
D 0.9301

20. What is 47.1093 rounded to the nearest hundredth?

F 47.1
G 47.2
H 47.109
J 47.11

Fractions

Reviewing Fractions

A **fraction** is a number that names part of a whole. Fractions are written in the form $\frac{a}{b}$, one number above another with a bar between them. The number on the bottom, the **denominator**, tells the number of equal parts in the whole. The number on the top, the **numerator**, tells the number of equal parts being described.

$$\frac{\text{numerator}}{\text{denominator}} = \frac{\text{parts described}}{\text{parts in whole}}$$

← $\frac{3}{5}$ are shaded →

Fraction of a whole

Fraction of a group

Types of Fractions: **Examples**

Proper fraction the numerator is less than the denominator $\frac{2}{3}, \frac{10}{17}, \frac{31}{50}$

Name for one the numerator and the denominator are the same $\frac{3}{3}, \frac{10}{10}, \frac{25}{25}$

Improper fraction the numerator is greater than the denominator $\frac{5}{2}, \frac{12}{7}, \frac{34}{21}$
An improper fraction names an amount greater than one.

Mixed number a whole number combined with a fraction $2\frac{1}{3}, 10\frac{1}{4}$

PRACTICE

Write the fraction for each situation below.

1. There are 100 cents in a dollar. What fraction of a dollar is 17 cents? ______

2. Brice had 3 energy bars. He ate half of one. How many energy bars remained uneaten? ______

3. Darlene has 59 pages to type. So far she has typed 23 pages. What fraction of the job is finished? ______

4. There are 12 inches in a foot. What fraction of a foot is 5 inches? ______

5. After spending $12, Mario still had $4 left. What fraction of his money had Mario spent? ______

6. There are 16 ounces in one pound. What mixed number represents 18 ounces expressed in pounds? ______

Comparing Fractions

The denominator of a fraction tells the number of equal parts in the whole. As the number of parts in the whole increases, the number in the denominator becomes greater and the size of each part becomes smaller. Sixths are smaller than thirds; eighths are smaller than fourths.

- If denominators are the same, compare numerators. The fraction with the greater number in the numerator is greater. The parts in both fractions are the same size, but there are more of those parts in the fraction with the greater number in the numerator.

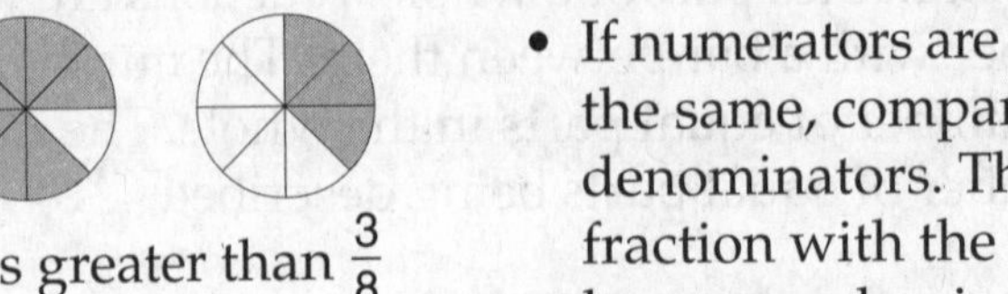
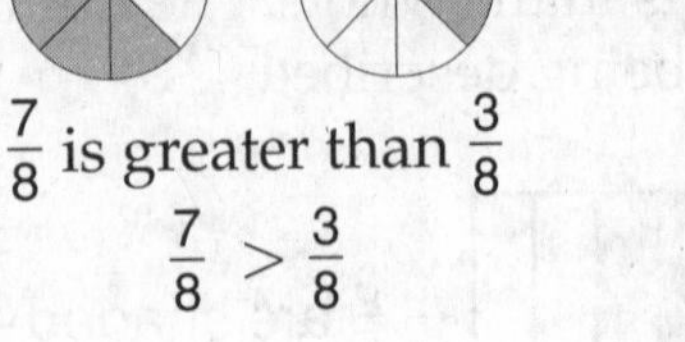

$\frac{7}{8}$ is greater than $\frac{3}{8}$

$\frac{7}{8} > \frac{3}{8}$

- If numerators are the same, compare denominators. The fraction with the lower number in the denominator is greater. Both fractions have the same number of parts, but each part is bigger in the fraction with the lower number in the denominator.

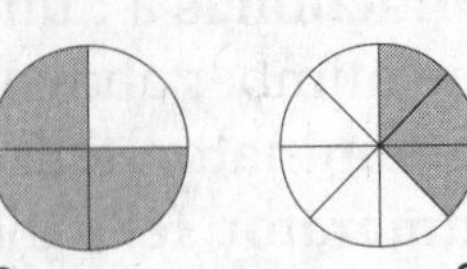

$\frac{3}{4}$ is greater than $\frac{3}{8}$

$\frac{3}{4} > \frac{3}{8}$

If *neither* the numerators *nor* the denominators are the same, the fractions are usually renamed to have a common denominator. (See pages 62–63.)

PRACTICE

Compare. Write $>$ (is greater than) or $<$ (is less than) between each pair of fractions. Draw a picture for each amount. The first one has been done for you.

1. $\frac{1}{3} > \frac{1}{6}$

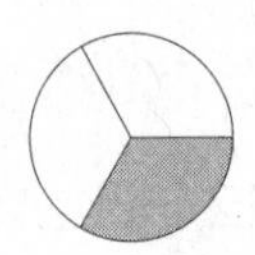
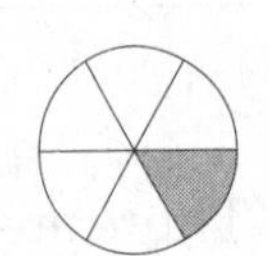

2. $\frac{1}{3}$ $\frac{1}{2}$

3. $\frac{1}{5}$ $\frac{1}{7}$

4. $\frac{6}{10}$ $\frac{9}{10}$

5. $\frac{3}{12}$ $\frac{3}{4}$

Rewrite each set of fractions in order from least to greatest.

6. $\frac{1}{7}$ $\frac{1}{5}$ $\frac{1}{9}$

7. $\frac{2}{3}$ $\frac{2}{5}$ $\frac{2}{9}$

8. $\frac{4}{8}$ $\frac{4}{7}$ $\frac{4}{9}$

9. $\frac{3}{7}$ $\frac{5}{7}$ $\frac{1}{7}$

10. $\frac{4}{5}$ $\frac{2}{5}$ $\frac{3}{5}$

11. $\frac{2}{9}$ $\frac{4}{9}$ $\frac{7}{9}$

Finding Equivalent Fractions

Equivalent fractions are equal in value, but they have different numerators and different denominators.

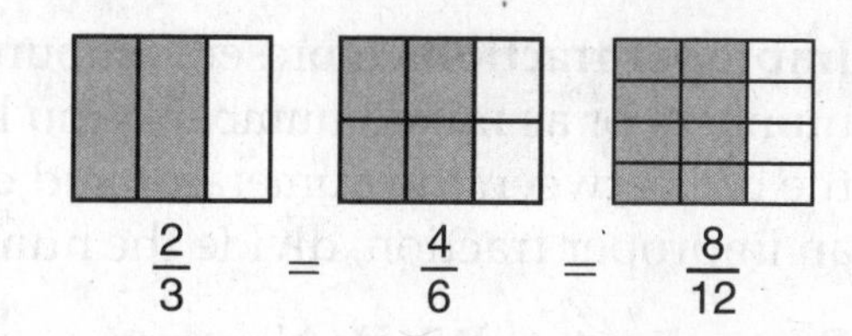

$$\frac{2}{3} = \frac{4}{6} = \frac{8}{12}$$

You can rename a fraction to an equivalent fraction in **higher terms**. To rename to higher terms, **multiply** both the numerator and the denominator by the same number or factor. The numbers in both numerator and denominator increase. The size of each part becomes smaller, so you need more of them to have an equal amount. (Think of pennies and nickels; you need five pennies to have the same value as one nickel.)

$$\frac{2}{3} = \frac{2 \times 2}{2 \times 3} = \frac{4}{6} \qquad \frac{2}{3} = \frac{3 \times 2}{3 \times 3} = \frac{6}{9}$$

$$\frac{2}{3} = \frac{4 \times 2}{4 \times 3} = \frac{8}{12} \qquad \frac{2}{3} = \frac{4}{6} = \frac{6}{9} = \frac{8}{12}$$

PRACTICE

Fill in the missing numbers to rename each fraction to higher terms.

1. $\frac{3}{7} = \frac{4 \times 3}{\square \times 7} = \frac{12}{28}$

2. $\frac{4}{5} = \frac{\square \times 4}{4 \times 5} = \frac{16}{\square}$

3. $\frac{2}{5} = \frac{\square \times 2}{\square \times 5} = \frac{12}{30}$

4. $\frac{1}{3} = \frac{5 \times 1}{\square \times 3} = \frac{\square}{\square}$

You can also rename to an equivalent fraction in **lower terms**. To rename to lower terms, **divide** both the numerator and the denominator by the same number or factor. The numbers in the numerator and denominator are decreased (lowered) to simpler numbers, so this is often called **reducing** or **simplifying** the fraction.

$$\frac{12}{30} = \frac{12 \div 2}{30 \div 2} = \frac{6}{15} \qquad \frac{6}{15} = \frac{6 \div 3}{15 \div 3} = \frac{2}{5}$$

$$\frac{12}{30} = \frac{6}{15} = \frac{2}{5}$$

When the only number that will evenly divide both the numerator and denominator is 1, the fraction is in its simplest terms or lowest terms.

PRACTICE

Fill in the missing numbers to rename each fraction to simplest terms.

5. $\frac{12}{14} = \frac{12 \div 2}{14 \div \square} = \frac{6}{7}$

6. $\frac{8}{20} = \frac{8 \div \square}{20 \div 4} = \frac{2}{\square}$

7. $\frac{9}{15} = \frac{9 \div \square}{15 \div \square} = \frac{3}{5}$

8. $\frac{6}{18} = \frac{6 \div \square}{18 \div \square} = \frac{\square}{3}$

9. $\frac{14}{35} = \frac{14 \div \square}{35 \div \square} = \frac{\square}{5}$

10. $\frac{9}{12} = \frac{9 \div \square}{12 \div \square} = \frac{3}{\square}$

Renaming Improper Fractions

Improper fractions represent amounts greater than 1. They can be renamed as whole numbers or as mixed numbers. You know that a fraction represents a division problem; the bar between the numerator and denominator represents a division sign. To rename an improper fraction, divide the numerator by the denominator.

Renaming to a Whole Number

- When the numerator and the denominator are the same, the fraction is a name for 1.
- When the numerator is a multiple of the denominator, the fraction represents a whole number greater than 1.

Examples

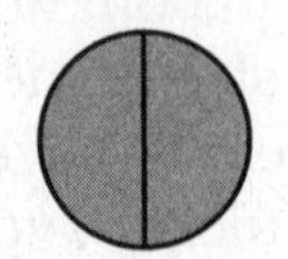

$\frac{2}{2} = 2 \div 2 = 1$

$\frac{4}{2} = 4 \div 2 = 2$

$\frac{4}{2} = 2$

Renaming to a Mixed Number

- When the numerator is not a multiple of the denominator, dividing will result in a whole number with a remainder. The remainder will become the numerator of the fraction for the mixed number, and the denominator of the improper fraction will be the denominator.

$\frac{7}{4} = 7 \div 4 = 1 \text{ R } 3 = 1\frac{3}{4}$

$\frac{7}{4} = 1\frac{3}{4}$

PRACTICE

Circle *W* if the fraction can be renamed to a whole number. Circle *M* if the fraction will be renamed to a mixed number.

1. $\frac{3}{3}$ W M
2. $\frac{5}{4}$ W M
3. $\frac{7}{6}$ W M
4. $\frac{33}{25}$ W M
5. $\frac{50}{50}$ W M
6. $\frac{12}{5}$ W M
7. $\frac{60}{6}$ W M
8. $\frac{18}{3}$ W M
9. $\frac{7}{4}$ W M
10. $\frac{15}{5}$ W M

Write each improper fraction as a whole number or as a mixed number. Reduce all proper fractions to simplest terms.

11. $\frac{4}{3}$ = ________
12. $\frac{6}{3}$ = ________
13. $\frac{5}{4}$ = ________
14. $\frac{50}{10}$ = ________
15. $\frac{95}{40}$ = ________
16. $\frac{12}{8}$ = ________
17. $\frac{7}{5}$ = ________
18. $\frac{30}{15}$ = ________
19. $\frac{9}{5}$ = ________
20. $\frac{18}{5}$ = ________

Moving Between Fractions and Decimals

The shaded portion of each figure shown at the right can be represented by either a fraction or a decimal.

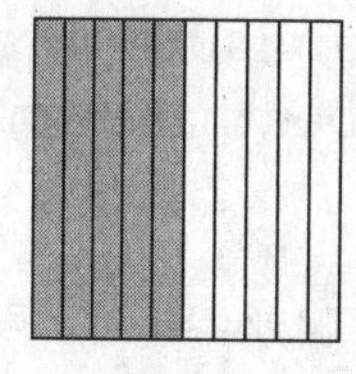

$\frac{1}{2} = 0.5$

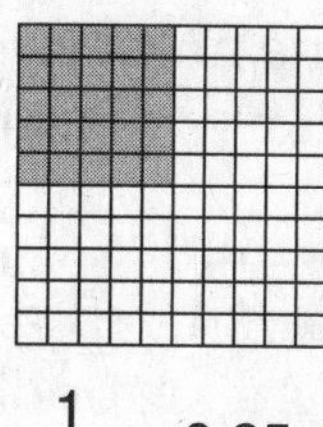

$\frac{1}{4} = 0.25$

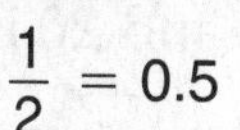

Any decimal can be written as a fraction with a denominator of 10 or 100 or 1,000, and so on. That fraction can usually be simplified.

To change a decimal to a fraction:

- Say or write the decimal in words.
- Use the place value you say as the denominator.
- Write the number you say in the numerator.
- Reduce the fraction to simplest terms.

Example

$0.2 = 2 \text{ tenths} = \frac{2}{10} = \frac{1}{5}$

$0.2 = \frac{1}{5}$

Any fraction can be renamed as a decimal. Divide the numerator by the denominator.

To change a fraction to a decimal:

- Read the fraction as a division problem.
- Divide the numerator by the denominator.

Example

$\frac{1}{4} = 1 \div 4 = 0.25$

$\frac{1}{4} = 0.25$

PRACTICE

Write each decimal as a fraction with a denominator of 10, 100, or 1,000. Then reduce the fraction to simplest terms.

1. 0.4 = ______
2. 0.45 = ______
3. 0.08 = ______
4. 0.125 = ______
5. 0.60 = ______
6. 3.5 = ______
7. 0.05 = ______
8. 0.75 = ______

Write each fraction as a decimal.

9. $\frac{1}{5} =$ ______
10. $\frac{3}{8} =$ ______
11. $\frac{3}{100} =$ ______
12. $\frac{8}{25} =$ ______
13. $\frac{9}{10} =$ ______
14. $\frac{7}{20} =$ ______
15. $\frac{6}{8} =$ ______

Using Divisibility Rules

One number is **divisible** by another if it can be evenly divided (no remainder) by the second number. A number that evenly divides another is a factor of the number it divides.

Examples

$20 \div 4 = 5$ There is no remainder.
4 is a factor of 20, and 20 is divisible by 4.

$20 \div 6 = 3 \text{ R } 2$ There is a remainder.
6 is not a factor of 20, and 20 is *not* divisible by 6.

These divisibility rules can help you determine whether one number is divisible by another.

Divisible by	Rule	Example
2	Even numbers—numbers with 0, 2, 4, 6, or 8 in the ones place—are divisible by 2.	976 is an even number. 976 is divisible by 2.
3	Add the digits. If the sum is divisible by 3, the number is divisible by 3.	402 → $4 + 0 + 2 = 6$ 6 is divisible by 3. 402 is divisible by 3.
4	If the last two digits of the number (the tens and ones digits) are divisible by 4, the number is divisible by 4.	The last two digits of 712 are 12. 12 is divisible by 4. 712 is divisible by 4.
5	If the number has 0 or 5 in the ones place, it is divisible by 5.	80 has 0 in the ones place. 345 has 5 in the ones place. Both 80 and 345 are divisible by 5.
6	If the number is even *and* it is divisible by 3, it is divisible by 6.	72 is an even number. $7 + 2 = 9$, and 9 is divisible by 3. 72 is divisible by 6.
8	If the last three digits of the number are divisible by 8, the number is divisible by 8.	The last three digits of 65,448 are 448. 448 is divisible by 8. 65,448 is divisible by 8.
9	If the sum of the digits is a multiple of 9, the number is divisible by 9.	24,534 → $2 + 4 + 5 + 3 + 4 = 18$ 18 is a multiple of 9. 24,534 is divisible by 9.
10	If the number has 0 in the ones place, it is divisible by 10.	2,380 has 0 in the ones place. 2,380 is divisible by 10.

PRACTICE

Circle the letter(s) of the correct answers. There is more than one correct answer for each question.

1. These numbers are divisible by both 3 and 9.

   A 327 **C** 891
   B 3,219 **D** 40,104

2. These numbers are divisible by 4.

   F 516 **H** 214
   G 412 **J** 84

3. These numbers are divisible by both 3 and 4.

   A 612 **C** 68
   B 232 **D** 324

4. These numbers are divisible by 2, 5, and 10.

   F 125 **H** 250
   G 100 **J** 70

Finding Common Factors to Simplify Fractions

A factor is a number that is multiplied to yield a product. Every whole number except the number 1 has at least two factors. An easy way to find the factors of a number is to list pairs that have that number as their product. Start with 1 times the number. Then try 2 times the number, 3 times the number, and so on. Work in order and list only pairs whose product is the number you are looking for. When you get to a pair that is already listed, you can stop.

Examples

Find the factors of 12.

$1 \times 12 = 12$ $2 \times 6 = 12$ $3 \times 4 = 12$ $4 \times 3 = 12$ Stop—we have 3 and 4.

List the factors in order from least to greatest.

The factors of 12 are 1, 2, 3, 4, 6, and 12.

Find the factors of 8.

$1 \times 8 = 8$ $2 \times 4 = 8$ 3 is not a factor $4 \times 2 = 8$ Stop—we have 2 and 4.

List the factors in order from least to greatest.

The factors of 8 are 1, 2, 4, and 8.

To simplify a fraction, divide the numerator and denominator by a **common factor**. The most efficient way to simplify is to divide by the **greatest common factor (GCF)**, which is the greatest number that is a factor of both the numerator and denominator.

Example Simplify $\frac{8}{12}$.

- List the factors of each number. 8: 1, 2, 4, 8 12: 1, 2, 3, 4, 6, 12
- Circle the common factors. The greatest common factor is 4.
- Divide numerator and denominator by the greatest common factor. $\frac{8}{12} = \frac{8 \div 4}{12 \div 4} = \frac{2}{3}$

PRACTICE

Find all of the factors of each number. List them in order from least to greatest.

1. factors of 18 __________
2. factors of 15 __________
3. factors of 20 __________
4. factors of 24 __________
5. factors of 28 __________
6. factors of 9 __________
7. factors of 8 __________
8. factors of 7 __________

Write the greatest common factor for the numerator and denominator of each fraction. Use the list of factors above to help. Then divide by that factor and simplify the fraction. The first one has been done for you.

9. $\frac{9}{18}$ GCF = 9 $\frac{9}{18} = \frac{1}{2}$
10. $\frac{15}{20}$ GCF = ____ $\frac{15}{20} =$ ____
11. $\frac{18}{24}$ GCF = ____ $\frac{18}{24} =$ ____
12. $\frac{8}{28}$ GCF = ____ $\frac{8}{28} =$ ____

Adding and Subtracting Fractions with Like Denominators

When you add, you put like units together; ones plus ones, tens plus tens, and so on. When you subtract, you remove like units; ones minus ones, tens minus tens, and so on. In fractions, the denominators identify the units—for example, fifths, eighths, or tenths. To add or subtract fractions, the fractions must have like (same) denominators.

To add or subtract fractions with like denominators:

- Write the denominator of the fractions in the denominator of the answer.
- Then add or subtract the numerators.
- Reduce to simplest terms.

Add. $\frac{3}{8} + \frac{1}{8} = \frac{4}{8} = \frac{1}{2}$

Subtract. $\frac{7}{10} - \frac{3}{10} = \frac{4}{10} = \frac{2}{5}$

To add or subtract mixed numbers with like denominators:

- Add or subtract the fractions first.
- Then add or subtract the whole numbers.
- Reduce to simplest terms.

Add.

$$\begin{array}{r} 1\frac{1}{8} \\ +\ 3\frac{5}{8} \\ \hline 4\frac{6}{8} \end{array} = 4\frac{3}{4}$$

Subtract.

$$\begin{array}{r} 3\frac{5}{6} \\ -\ 1\frac{2}{6} \\ \hline 2\frac{3}{6} \end{array} = 2\frac{1}{2}$$

When the answer has an improper fraction, rename.

- Combine whole numbers.
- Reduce fraction to simplest terms.

$$\begin{array}{r} 3\frac{3}{4} \\ +\ 2\frac{3}{4} \\ \hline 5\frac{6}{4} \end{array} = 5 + 1 + \frac{2}{4} = 6\frac{2}{4} = 6\frac{1}{2}$$

PRACTICE

Add. Simplify answers.

1. $\frac{2}{9} + \frac{5}{9} =$ ______
2. $\frac{3}{10} + \frac{7}{10} =$ ______
3. $\frac{4}{9} + \frac{8}{9} =$ ______
4. $\frac{8}{15} + \frac{2}{15} =$ ______
5. $\frac{5}{12} + \frac{1}{12} + \frac{7}{12} =$ ______
6. $1\frac{1}{4} + \frac{1}{4} =$ ______
7. $3\frac{1}{3} + 2\frac{1}{3} =$ ______
8. $3\frac{3}{8} + 2\frac{5}{8} =$ ______
9. $2\frac{7}{16} + 1\frac{11}{16} + \frac{3}{16} =$ ______

Subtract. Simplify answers.

10. $\frac{9}{15} - \frac{4}{15} =$ ______
11. $\frac{4}{12} - \frac{3}{12} =$ ______
12. $\frac{5}{6} - \frac{1}{6} =$ ______
13. $\frac{7}{8} - \frac{3}{8} =$ ______
14. $1\frac{7}{10} - \frac{5}{10} =$ ______
15. $2\frac{2}{9} - \frac{2}{9} =$ ______
16. $3\frac{3}{7} - 2\frac{1}{7} =$ ______
17. $10\frac{3}{8} - 6\frac{1}{8} =$ ______
18. $5\frac{7}{10} - 1\frac{3}{10} =$ ______
19. $7\frac{9}{15} - 2\frac{9}{15} =$ ______

Regrouping to Subtract Fractions

You may need to regroup a whole number to a fraction before you can subtract. This process is similar to what you do when you subtract with whole numbers, but with fractions you will regroup to a name for one. The fraction you are working with determines what name for one to use.

Examples Regroup to get fourths. Regroup to get more halves.

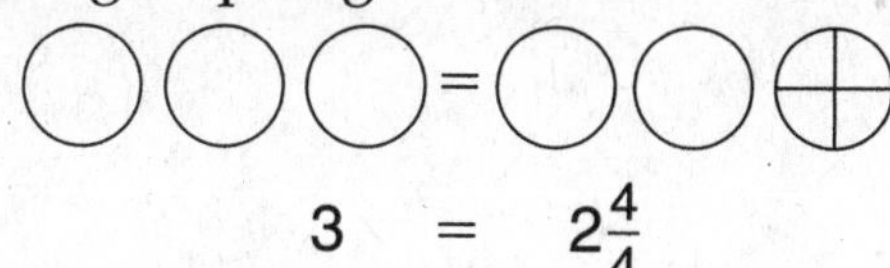

$3 = 2\frac{4}{4}$

$2\frac{1}{2} = 1 + \frac{2}{2} + \frac{1}{2} = 1\frac{3}{2}$

PRACTICE

Fill in the missing numbers to name each whole number or mixed number to an equivalent amount.

1. $1 = \frac{6}{\square}$
2. $1 = \frac{\square}{8}$
3. $2 = 1\frac{5}{\square}$
4. $5 = 4\frac{\square}{7}$
5. $1\frac{1}{10} = \frac{\square}{10}$
6. $2\frac{1}{4} = 1\frac{\square}{4}$
7. $9\frac{2}{3} = \square\frac{5}{3}$
8. $4\frac{3}{4} = 3\frac{\square}{\square}$

You may need to regroup a whole number or a mixed number before you can subtract.

Examples

$3 - 1\frac{1}{6}$ Regroup 3 to get sixths. Think: $3 = 2\frac{6}{6}$

Then subtract.
Simplify the answer if needed.

$$\begin{array}{r} 3 = 2\frac{6}{6} \\ -1\frac{1}{6} = 1\frac{1}{6} \\ \hline 1\frac{5}{6} \end{array}$$

$4\frac{1}{8} - 2\frac{5}{8}$ Regroup $4\frac{1}{8}$.
Think:
$4\frac{1}{8} = 3 + 1 + \frac{1}{8}$
$= 3 + \frac{8}{8} + \frac{1}{8} = 3\frac{9}{8}$

Then subtract.
Simplify the answer if needed.

$$\begin{array}{r} 4\frac{1}{8} = 3\frac{9}{8} \\ -2\frac{5}{8} = 2\frac{5}{8} \\ \hline 1\frac{4}{8} = 1\frac{1}{2} \end{array}$$

PRACTICE

Subtract. Make sure each difference is in simplest terms.

9. $1 - \frac{4}{5}$
10. $2 - \frac{3}{4}$
11. $5\frac{3}{7} - 2\frac{5}{7}$
12. $2\frac{3}{8} - \frac{4}{8}$
13. $4\frac{1}{3} - 3\frac{2}{3}$
14. $5\frac{1}{6} - 2\frac{5}{6}$
15. $12\frac{2}{5} - 3\frac{4}{5}$
16. $1\frac{7}{30} - \frac{11}{30}$

Finding Common Denominators

Fractions with unlike denominators are renamed to have a **common denominator** so they can be added, subtracted, or compared. Using the **least common denominator (LCD)** can make your work easier because you will have smaller numbers to work with, and answers for addition and subtraction will already be in simplest form.

To find the least common denominator, look for the **least common multiple (LCM)** of the numbers in the denominators.

Examples

Find the least common multiple of 6 and 8.

- Write multiples of each number.
- Circle common multiples.
- Select the least common multiple.

6: 6, 12, 18, 24, 30, 36, 42, 48...

8: 8, 16, 24, 32, 40, 48...

The least common multiple is 24.

The least common multiple of 6 and 8 is 24.

Find the least common multiple of 10 and 25.

- Write multiples of each number.
- Circle the common multiples.
- Select the least common multiple.

10: 10, 20, 30, 40, 50, 60, 70, 80, 90, 100...

25: 25, 50, 75, 100...

The least common multiple is 50.

The least common multiple of 10 and 25 is 50.

PRACTICE

List the multiples for each denominator. Then find the least common multiple for each pair of fractions. That number is the least common denominator for the pair.

1. $\frac{1}{5}$ and $\frac{2}{3}$ Multiples of 5 ________

   Multiples of 3 ________

   LCM ________

2. $\frac{3}{4}$ and $\frac{2}{7}$ Multiples of 4 ________

   Multiples of 7 ________

   LCM ________

3. $\frac{4}{9}$ and $\frac{1}{6}$ Multiples of 9 ________

   Multiples of 6 ________

   LCM ________

4. $\frac{3}{8}$ and $\frac{5}{12}$ Multiples of 8 ________

   Multiples of 12 ________

   LCM ________

Comparing and Ordering Unlike Fractions

When fractions have like denominators, you can compare them by looking at their numerators. The fraction with the greater numerator is greater. $\frac{2}{3} > \frac{1}{3}$

When fractions have like numerators, look at their denominators to compare them. The fraction with the lesser denominator is greater. $\frac{1}{2} > \frac{1}{8}$

If fractions have *neither* like numerators *nor* like denominators, there are different methods to use to compare them. One method is to rename the fractions to a common denominator and then compare the numerators. You can use the least common multiple as the common denominator.

Example Compare $\frac{3}{5}$ and $\frac{7}{9}$.

- Write multiples of each denominator.
  9: 9, 18, 27, 36, 45, ...
  5: 5, 10, 15, 20, 25, 30, 35, 40, 45, ...
- Find the least common multiple.
  The least common multiple is 45.
- Rename both fractions to a common denominator. Use the least common multiple as the denominator.

$$\frac{3}{5} = \frac{9 \times 3}{9 \times 5} = \frac{27}{45} \qquad \frac{7}{9} = \frac{5 \times 7}{5 \times 9} = \frac{35}{45}$$

$\frac{27}{45} < \frac{35}{45}$ so $\frac{3}{5} < \frac{7}{9}$

PRACTICE

Write >, <, or = to compare each pair of fractions.

1. $\frac{4}{7}$ $\frac{2}{3}$
2. $\frac{7}{10}$ $\frac{4}{5}$
3. $\frac{5}{8}$ $\frac{7}{12}$
4. $\frac{2}{3}$ $\frac{3}{5}$
5. $\frac{3}{4}$ $\frac{6}{8}$
6. $\frac{7}{9}$ $\frac{5}{6}$

Write each set of fractions in order from least to greatest.

7. $\frac{7}{12}, \frac{2}{3}, \frac{3}{4}$
8. $\frac{5}{12}, \frac{1}{3}, \frac{7}{18}$
9. $\frac{3}{4}, \frac{7}{8}, \frac{1}{2}$
10. $\frac{2}{5}, \frac{1}{4}, \frac{3}{10}$
11. $\frac{5}{8}, \frac{5}{6}, \frac{7}{9}$
12. $\frac{9}{20}, \frac{3}{5}, \frac{7}{10}$

Adding and Subtracting Fractions with Unlike Denominators

To add or subtract fractions with unlike denominators, first rewrite them to have a common denominator. Try to find the least common denominator.

Examples

Add. $\frac{3}{5} + \frac{2}{3}$

$$\frac{3}{5} = \frac{9}{15}$$
$$+\ \frac{2}{3} = \frac{10}{15}$$
$$\frac{19}{15} = 1\frac{4}{15}$$

Subtract. $\frac{11}{14} - \frac{3}{7}$

$$\frac{11}{14} = \frac{11}{14}$$
$$-\ \frac{3}{7} = \frac{6}{14}$$
$$\frac{5}{14}$$

PRACTICE

Rename fractions to have a common denominator. Then add or subtract as indicated. Make sure answers are in simplest terms.

1. $\frac{3}{2} + \frac{1}{5}$

2. $\frac{5}{7} + \frac{3}{4}$

3. $\frac{1}{2} + \frac{2}{3}$

4. $\frac{1}{4} + \frac{3}{20}$

5. $\frac{6}{7} - \frac{1}{4}$

6. $\frac{5}{9} - \frac{1}{3}$

7. $\frac{3}{5} - \frac{1}{10}$

8. $\frac{11}{24} - \frac{3}{8}$

9. $\frac{1}{3} + \frac{8}{9} =$ ______

10. $\frac{4}{9} - \frac{1}{5} =$ ______

11. $\frac{2}{7} + \frac{1}{5} =$ ______

12. $\frac{5}{8} - \frac{1}{4} =$ ______

13. $\frac{3}{8} + \frac{1}{6} =$ ______

14. $\frac{8}{20} - \frac{2}{5} =$ ______

15. $\frac{7}{8} + \frac{2}{3} =$ ______

16. $\frac{7}{12} - \frac{1}{3} =$ ______

17. $\frac{3}{4} + \frac{5}{12} =$ ______

Multiplying Fractions

- To multiply fractions, multiply the numerators together. Then multiply the denominators together. Simplify answers. Remember, in math, the word *of* means multiply.
- To multiply a fraction and a whole number, rewrite the whole number as an improper fraction with a denominator of 1. Then multiply the numerators and multiply the denominators. Simplify answers.

Examples

Find $\frac{2}{3}$ of $\frac{3}{6}$.

$$\frac{2}{3} \times \frac{3}{6} = \frac{2 \times 3}{3 \times 6} = \frac{6}{18}$$

$$\frac{6}{18} = \frac{6 \div 6}{18 \div 6} = \frac{1}{3}$$

$$\frac{2}{3} \times \frac{3}{6} = \frac{1}{3}$$

Find $\frac{2}{3}$ of 6.

$$\frac{2}{3} \times 6 = \frac{2}{3} \times \frac{6}{1}$$

$$= \frac{2 \times 6}{3 \times 1} = \frac{12}{3} = 4$$

$$\frac{2}{3} \times 6 = 4$$

PRACTICE

Multiply. Reduce your answers to simplest terms.

1. $\frac{1}{4} \times \frac{3}{5} =$
2. $\frac{1}{3} \times \frac{2}{5} =$
3. $\frac{2}{3} \times \frac{2}{3} =$
4. $\frac{1}{5} \times \frac{3}{5} =$
5. $9 \times \frac{2}{6} =$
6. $\frac{3}{5} \times \frac{2}{3} =$
7. $\frac{1}{8} \times 3 =$
8. $\frac{4}{9} \times \frac{1}{5} =$
9. $7 \times \frac{1}{5} =$
10. $\frac{1}{2} \times \frac{1}{2} \times \frac{2}{3} =$
11. $\frac{1}{2} \times \frac{2}{5} \times \frac{3}{4} =$
12. $\frac{1}{7} \times \frac{3}{5} \times 3 =$
13. $\frac{1}{3} \times \frac{2}{3} \times 5 =$
14. $\frac{3}{7} \times 4 =$
15. What is $\frac{4}{5}$ of 10?
16. What is $\frac{2}{3}$ of 15?
17. What is $\frac{1}{5}$ of 6?
18. What is $\frac{3}{8}$ of 40?
19. Elaine is making half a recipe for rice. The recipe calls for $\frac{1}{4}$ cup of butter. What is $\frac{1}{2}$ of $\frac{1}{4}$ cup?
20. At Don's office, $\frac{2}{9}$ of the people smoke. There are 45 workers in the office. How many of them smoke?

Canceling Before Multiplying Fractions

Canceling is a shortcut that lets you simplify fractions *before* you multiply. Canceling allows you to multiply smaller numbers and eliminates the need to simplify the product.

To cancel, look for a common factor of the numerator and the denominator. Divide both the numerator and the denominator by the common factor, and then multiply as usual. Be sure you multiply the numbers that resulted from the canceling.

Example Multiply. $\frac{3}{4} \times \frac{2}{3} \times \frac{1}{2}$

- Look for a common factor in the numerator and the denominator. There are *two* common factors, 2 and 3. $\frac{3 \times \cancel{2}^{1} \times 1}{4 \times 3 \times \cancel{2}_{1}}$
- Divide numerator and denominator by 2.
- Divide numerator and denominator by 3. $\frac{\cancel{3}^{1} \times \cancel{2}^{1} \times 1}{4 \times \cancel{3}_{1} \times \cancel{2}_{1}} = \frac{1}{4}$
- Multiply as usual.

$$\frac{3}{4} \times \frac{2}{3} \times \frac{1}{2} = \frac{1}{4}$$

PRACTICE

Simplify the problems below by canceling. Then multiply.

1. $\frac{2}{9} \times \frac{3}{5} =$
2. $\frac{1}{6} \times \frac{2}{3} =$
3. $\frac{3}{10} \times \frac{5}{9} =$
4. $\frac{3}{4} \times \frac{1}{9} \times \frac{2}{5} =$
5. $10 \times \frac{4}{5} =$
6. $\frac{2}{3} \times \frac{3}{4} \times \frac{4}{5} =$
7. $6 \times \frac{1}{10} =$
8. $3 \times \frac{1}{3} \times \frac{1}{5} =$
9. $\frac{2}{7} \times \frac{1}{2} =$
10. $\frac{6}{7} \times \frac{1}{2} =$
11. What is $\frac{1}{3}$ of 33?
12. How much meat is in $\frac{2}{3}$ of $\frac{3}{4}$ of a pound of hamburger?
13. What is $\frac{4}{5}$ of $\frac{1}{2}$?
14. What is $\frac{1}{4}$ of 64?
15. What is $\frac{5}{6}$ of $\frac{3}{4}$?

Finding Reciprocals

You use **reciprocals** to divide fractions. A reciprocal is one of two numbers whose product is 1.

- To find the reciprocal of a fraction, switch the numerator and the denominator.

Examples

$\frac{2}{3} \times \frac{3}{2} = \frac{6}{6} = 1$ $\frac{3}{2}$ is the reciprocal of $\frac{2}{3}$, and $\frac{2}{3}$ is the reciprocal of $\frac{3}{2}$.

$\frac{5}{8} \times \frac{8}{5} = \frac{40}{40} = 1$ $\frac{5}{8}$ and $\frac{8}{5}$ are reciprocals of each other.

- To find the reciprocal of a whole number, write the number over a denominator of 1. Then switch the numerator and the denominator.

Example

$4 = \frac{4}{1}$ The reciprocal of $\frac{4}{1}$ is $\frac{1}{4}$, so the reciprocal of 4 is $\frac{1}{4}$.

- To find the reciprocal of a mixed number, change the mixed number to an improper fraction and find the reciprocal of the improper fraction. Do not simplify.

Example

$2\frac{3}{5} = \frac{13}{5}$ The reciprocal of $\frac{13}{5}$ is $\frac{5}{13}$, so the reciprocal of $2\frac{3}{5}$ is $\frac{5}{13}$.

PRACTICE

Find the reciprocal of each number. Do *not* simplify.

1. $\frac{4}{5}$ ______
2. 7 ______
3. $1\frac{1}{3}$ ______
4. $3\frac{1}{2}$ ______
5. $\frac{5}{12}$ ______
6. $\frac{1}{9}$ ______
7. 6 ______
8. $\frac{2}{7}$ ______
9. $4\frac{1}{2}$ ______
10. $\frac{6}{9}$ ______
11. $\frac{1}{8}$ ______
12. 15 ______
13. 2 ______
14. $1\frac{1}{8}$ ______
15. $\frac{9}{10}$ ______

Dividing Fractions

To divide with fractions, you rewrite a division problem as a multiplication problem. You will need to change the divisor (the number you are dividing by) to its reciprocal.

Examples

Find $\frac{3}{5} \div \frac{2}{3}$.

- Rewrite the division problem as a multiplication problem.
  - Change the division sign to a multiplication sign.
  - Write the reciprocal of the divisor.
- Multiply.

$\frac{3}{5} \div \frac{2}{3} = \frac{3}{5} \times \frac{3}{2}$

$\frac{3}{5} \times \frac{3}{2} = \frac{9}{10}$

$\frac{3}{5} \div \frac{2}{3} = \frac{9}{10}$

Find $\frac{5}{6} \div \frac{2}{3}$.

- Rewrite the division problem as a multiplication problem.
  - Change the division sign to a multiplication sign.
  - Write the reciprocal of the divisor.
- Multiply. Simplify the answer.

$\frac{5}{6} \div \frac{2}{3} = \frac{5}{6} \times \frac{3}{2}$

$\frac{5}{\not{6}_2} \times \frac{\not{3}^1}{2} = \frac{5}{4} = 1\frac{1}{4}$

$\frac{5}{6} \div \frac{2}{3} = 1\frac{1}{4}$

PRACTICE

Write the reciprocal to complete changing each division problem to a multiplication problem. Then solve. Write answers in simplest terms.

1. $\frac{3}{4} \div \frac{1}{2} = \frac{3}{4} \times \frac{\square}{\square} =$

2. $\frac{3}{5} \div \frac{2}{5} = \frac{3}{5} \times \frac{\square}{\square} =$

3. $\frac{4}{9} \div \frac{2}{3} = \frac{4}{9} \times \frac{\square}{\square} =$

4. $\frac{2}{3} \div \frac{5}{4} = \frac{2}{3} \times \frac{\square}{\square} =$

Divide. Write all answers in simplest terms.

5. $\frac{9}{10} \div \frac{3}{8} =$

6. $\frac{1}{4} \div \frac{4}{7} =$

7. $\frac{1}{2} \div \frac{1}{6} =$

8. $\frac{4}{5} \div \frac{3}{8} =$

9. $\frac{7}{8} \div \frac{2}{3} =$

10. $\frac{5}{6} \div \frac{5}{6} =$

11. $\frac{2}{9} \div \frac{3}{6} =$

12. $\frac{5}{7} \div \frac{1}{14} =$

13. $\frac{7}{12} \div \frac{1}{2} =$

14. $\frac{1}{2} \div \frac{7}{12} =$

15. $\frac{9}{10} \div \frac{3}{4} =$

16. $\frac{7}{25} \div \frac{1}{5} =$

Dividing with Whole Numbers or Mixed Numbers

When dividing with fractions, both the dividend (number being divided) and the divisor (number you are dividing by) must be written in fraction form.

- If one of the numbers is a whole number, write the whole number as a fraction with a denominator of 1. Then divide as usual.
- If one of the numbers is a mixed number, write the mixed number as an improper fraction. Then divide as usual.

You can check the division with multiplication. Multiply the answer by the divisor.

Examples

Find $4 \div \frac{7}{8}$.

$4 \div \frac{7}{8} = \frac{4}{1} \div \frac{7}{8}$

$\frac{4}{1} \div \frac{7}{8} = \frac{4}{1} \times \frac{8}{7} = \frac{32}{7} = 4\frac{4}{7}$

$4 \div \frac{7}{8} = 4\frac{4}{7}$

Check: $4\frac{4}{7} \times \frac{7}{8} = \frac{\cancel{32}^{4}}{\cancel{7}_{1}} \times \frac{\cancel{7}^{1}}{\cancel{8}_{1}} = 4$

Find $1\frac{2}{3} \div \frac{5}{6}$.

$1\frac{2}{3} \div \frac{5}{6} = \frac{5}{3} \div \frac{5}{6}$

$\frac{5}{3} \div \frac{5}{6} = \frac{\cancel{5}^{1}}{\cancel{3}_{1}} \times \frac{\cancel{6}^{2}}{\cancel{5}_{1}} = 2$

$1\frac{2}{3} \div \frac{5}{6} = 2$

Check: $\frac{2}{1} \times \frac{5}{6} = \frac{10}{6} = 1\frac{4}{6} = 1\frac{2}{3}$

PRACTICE

Find each quotient. Express each answer in simplest terms. Use multiplication to check your answers.

1. $12 \div \frac{2}{3} =$ ________
2. $6 \div \frac{3}{4} =$ ________
3. $9 \div \frac{3}{4} =$ ________
4. $\frac{4}{7} \div 4 =$ ________
5. $\frac{1}{3} \div 2 =$ ________
6. $\frac{2}{3} \div 15 =$ ________
7. $\frac{3}{5} \div 9 =$ ________
8. $\frac{4}{7} \div 12 =$ ________
9. $3\frac{3}{4} \div \frac{1}{2} =$ ________
10. $4 \div 1\frac{1}{2} =$ ________
11. $2\frac{2}{3} \div 4 =$ ________
12. $2\frac{2}{5} \div \frac{3}{10} =$ ________
13. Wayne is putting in 5 yards of fencing. He wants a fence post every $\frac{1}{2}$ yard. How many $\frac{1}{2}$-yard sections will he have? ________
14. Greta has 4 cups of juice. She is dividing it into $\frac{1}{2}$-cup portions. How many portions will she have? ________

Using Fractions to Solve Word Problems

Circle the correct answer for each problem.

1. Claude has saved $300 toward a leather couch he wants to buy. The couch costs $1,200. What fraction of the cost of the couch has he saved?

   A $\frac{1}{2}$ C $\frac{1}{3}$

   B $\frac{1}{4}$ D $\frac{1}{5}$

   E None of these

2. The Good Deeds Club raised $450 selling raffle tickets. One person bought $150 of raffle tickets. What fraction of the total amount raised did that person's tickets represent?

   F $\frac{1}{2}$ H $\frac{1}{3}$

   G $\frac{1}{4}$ J $\frac{1}{5}$

   K None of these

3. A $36 shirt is on sale for $\frac{1}{3}$ off. What is the sale price of the shirt?

   A $12 C $16

   B $20 D $24

   E None of these

4. Ron has $\frac{1}{2}$ of a tank of gas. If he puts another $\frac{1}{3}$ of a tank of gas into his car, what fraction of the tank will be filled?

   F $\frac{1}{5}$ H $\frac{1}{6}$

   G $\frac{5}{6}$ J $\frac{2}{3}$

   K None of these

Solve each word problem. Write all answers in simplest terms.

5. The label on a carton of pasta salad says the carton contains 12 cups of salad. If each serving is $\frac{2}{3}$ of a cup, how many servings of salad does the carton hold?

   ____

6. Leanne is trying to choose between two party dresses that originally cost $120. One dress is on sale for $\frac{1}{4}$ off, and the price of the other has been reduced by $40. What is the price of the less expensive dress?

   ____

7. Arun decided to walk $7\frac{1}{3}$ miles to his friend's house. After he had gone $3\frac{5}{8}$ miles, he stopped to rest. How much farther does he have to walk to reach the friend's house?

   ____

8. Ricky is getting married. Six members of his work crew are buying him a gift that costs $126, so each will pay $\frac{1}{6}$ of the cost. What is $\frac{1}{6}$ of $126?

   ____

9. Tim the baker uses $2\frac{1}{3}$ cups of flour for each chocolate cake he bakes. He plans to bake 9 chocolate cakes today. How many cups of flour will he use for the cakes?

   ____

10. You need $\frac{1}{4}$ of a yard of cloth to make a stuffed animal. The store has only $\frac{1}{3}$ of a yard of the cloth you want. Explain whether or not this will be enough cloth.

Fractions Skills Checkup

Circle the letter of the correct answer to each problem. Reduce all fractions to simplest terms.

1. $1\frac{3}{5} + \frac{2}{5} =$

   A $1\frac{1}{5}$ C $2\frac{1}{5}$
   B 2 D $2\frac{4}{5}$
   E None of these

2. $\frac{7}{10} - \frac{3}{10} =$

   F $\frac{3}{5}$ H $\frac{2}{3}$
   G $\frac{1}{5}$ J $\frac{2}{5}$
   K None of these

3. $5 \times \frac{2}{3} =$

   A $1\frac{1}{3}$ C $3\frac{1}{3}$
   B $2\frac{1}{3}$ D $\frac{2}{15}$
   E None of these

4. $\frac{3}{5} \div \frac{2}{5} =$

   F $\frac{2}{3}$ H $\frac{6}{25}$
   G $\frac{11}{2}$ J 1
   K None of these

5. $\frac{1}{3} \div 9 =$

   A 3 C $\frac{1}{27}$
   B 27 D $\frac{1}{9}$
   E None of these

6. $1\frac{3}{4} \div \frac{3}{7} =$

   F $18\frac{1}{3}$ H $\frac{3}{4}$
   G $\frac{4}{7}$ J $2\frac{1}{7}$
   K None of these

Use the following information to answer Numbers 7–10.

A school band has 480 boxes of popcorn to deliver. The booster club has offered to deliver $\frac{1}{3}$ of the boxes. The cheerleaders will deliver $\frac{1}{4}$ of the boxes.

7. How many boxes will the booster club deliver?

   A 120 C 160
   B 140 D 180

8. What fraction of the boxes will the members of the band deliver?

   F $\frac{5}{7}$ H $\frac{5}{12}$
   G $\frac{7}{12}$ J $\frac{2}{7}$

9. How many more boxes will be delivered by boosters than by cheerleaders?

   A 20 C 60
   B 40 D 80

10. Each box of popcorn weighs $\frac{1}{2}$ pound. Which of these number sentences could be used to figure out how many pounds of popcorn the band sold?

    F $480 \times 2 = \square$
    G $480 \div \frac{1}{2} = \square$
    H $480 \times \frac{1}{2} = \square$
    J $1{,}480 \times \frac{1}{2} = \square$

Fractions Skills Checkup (continued)

11. $\begin{array}{r} 9 \\ -\frac{1}{3} \\ \hline \end{array}$

A $9\frac{2}{3}$ C $8\frac{1}{3}$

B $9\frac{1}{3}$ D $8\frac{2}{3}$

E None of these

12. $2\frac{3}{5} + 5\frac{4}{5} =$

F $8\frac{2}{5}$ H $8\frac{1}{5}$

G $7\frac{5}{6}$ J $7\frac{2}{5}$

K None of these

13. $20 \div \frac{1}{2} =$

A 10 C 30

B $\frac{1}{40}$ D 40

E None of these

14. $\frac{2}{5} + \frac{3}{25} =$

F 1 H $\frac{13}{25}$

G $\frac{1}{5}$ J $\frac{6}{25}$

K None of these

15. $\frac{2}{3} - \frac{1}{6} =$

A $\frac{1}{6}$ C $\frac{1}{2}$

B $\frac{1}{3}$ D $\frac{2}{3}$

E None of these

16. Which fraction is equivalent to $\frac{3}{4}$?

A $\frac{6}{8}$ C $\frac{5}{6}$

B $\frac{1}{3}$ D $\frac{4}{3}$

17. Which set of fractions is in order from least to greatest?

F $\frac{1}{3}$ $\frac{1}{5}$ $\frac{1}{10}$ $\frac{1}{12}$

G $\frac{1}{12}$ $\frac{1}{3}$ $\frac{1}{5}$ $\frac{1}{10}$

H $\frac{1}{12}$ $\frac{1}{10}$ $\frac{1}{5}$ $\frac{1}{3}$

J $\frac{1}{10}$ $\frac{1}{5}$ $\frac{1}{3}$ $\frac{1}{12}$

18. Which number sentence is true?

A $\frac{3}{4} > 1$ C $\frac{9}{8} > 1$

B $\frac{0}{5} = 1$ D $\frac{3}{7} > 1$

19. Which fraction is equal to 0.5?

F $\frac{1}{3}$ H $\frac{3}{4}$

G $\frac{1}{2}$ J $\frac{1}{5}$

20. Seven city council members voted to rezone Oak Avenue. The other 8 council members voted against rezoning it. What fraction of the council voted to rezone Oak Avenue?

A $\frac{7}{8}$ C $\frac{7}{15}$

B $\frac{8}{7}$ D $\frac{15}{7}$

Integers

Understanding Integers

Until now we have worked with **positive numbers**, numbers that are greater than zero. Each positive number has an opposite **negative number**.

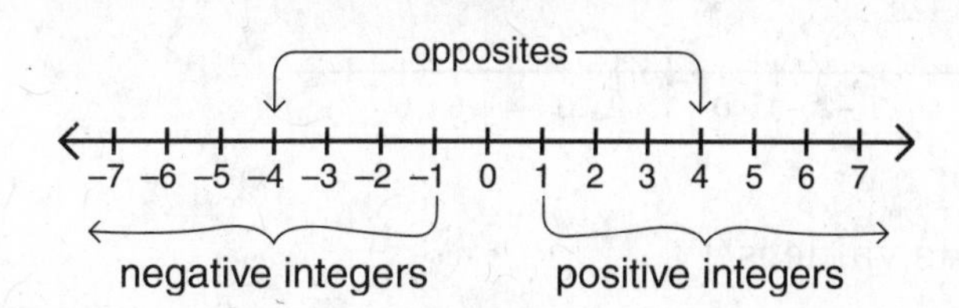

You are probably familiar with negative numbers being used to describe temperatures below zero. Negative numbers are identified by placing a minus sign in front of them. For example, 2 degrees below zero is written as $-2°$. A positive number may have a plus sign in front of it, but if there is no sign, the number is understood to be positive.

Positive whole numbers and their opposite negative whole numbers, together with zero, are called **integers**.

- As you move to the left on the number line, the value of each number decreases.
  -7 is to the left of -5, so -7 is less than -5. $-7 < -5$
  You can think about it this way: $-7°$ is colder than $-5°$, or, if you spend 7 dollars, you will have less money left than if you spend only 5 dollars.
- As you move to the right on the number line, the value increases.
  1 is to the right of -8, so 1 is greater than -8. $1 > -8$

PRACTICE

Compare the numbers in each pair. Write $<$, $>$, or $=$.

1. -5 ○ 3
2. -2 ○ -1
3. 0 ○ -6
4. $\frac{1}{2}$ ○ $-\frac{3}{4}$
5. -7 ○ -3
6. -1 ○ 6
7. -0.02 ○ -0.025
8. -0.5 ○ $-\frac{1}{2}$

Write the numbers in each set in order from least to greatest.

9. $5, -12, 3, 0$
10. $12, 4, -52, -13$
11. $-\frac{1}{2}, 3, -1.5, \frac{3}{4}$

Working with Absolute Value

The **absolute value** of a number is its distance from 0 on the number line.

Example
The number 4 is four spaces from 0. It has an absolute value of 4.

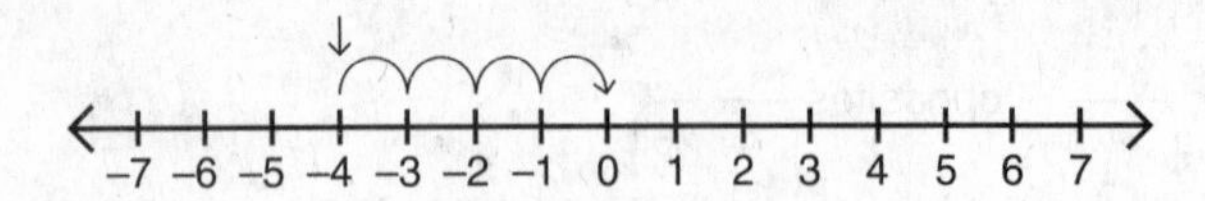

- The symbol for absolute value is | |.

  The absolute value of 4 is written $|4|$.

  The absolute value of −20 is written $|-20|$.

- A number's absolute value is always a positive number.

Examples

$|12| = 12$ $|-3| = 3$ $\left|-\frac{2}{3}\right| = \frac{2}{3}$ $|0.8| = 0.8$

PRACTICE

Write the absolute value.

1. $|3.1| =$ ______
2. $|-3| =$ ______
3. $|5| =$ ______
4. $|-5| =$ ______
5. $|-6| =$ ______
6. $|-0.25| =$ ______
7. $|32| =$ ______
8. $|0| =$ ______

Compare the values in each pair. Write <, >, or =.

9. $|32|$ ○ $|-32|$
10. $|-12|$ ○ $|-15|$
11. $\left|\frac{1}{4}\right|$ ○ $|-0.5|$
12. $|-5|$ ○ $|5|$
13. $|22|$ ○ $|-6|$
14. $|3|$ ○ $|-8|$
15. $|16.1|$ ○ $|-20|$
16. $|1|$ ○ $|1.0|$
17. $|-4|$ ○ 4
18. 0 ○ $|-2|$
19. $|3|$ ○ −9
20. $|12|$ ○ 15
21. −6 ○ $|-10|$
22. $|0.4|$ ○ 0.35
23. $|-6|$ ○ 6
24. 3 ○ $|-1|$

Adding with Integers

You can use absolute values to find the sums of positive and negative numbers.

- When you add two numbers with the same sign (both positive or both negative), find the sum of their absolute values. Then give the sum their common sign.

Examples Add $3 + 5$.
$|3| + |5| = 3 + 5 = 8$
Both addends are positive.
The sum is positive.
$3 + 5 = 8$
$(+) + (+) = (+)$

Add $^{-}3 + {}^{-}5$.
$|{}^{-}3| + |{}^{-}5| = 3 + 5 = 8$
Both addends are negative.
The sum is negative.
$^{-}3 + {}^{-}5 = {}^{-}8$
$(-) + (-) = (-)$

- When you add two numbers with unlike signs, **find the difference in their absolute values**. Then give that difference the sign of the addend with the greater absolute value.

Examples Add $2 + {}^{-}2$.
$|2| - |{}^{-}2| = 2 - 2 = 0$
Zero is not positive or negative.
$^{*}2 + {}^{-}2 = 0$

Add $4 + {}^{-}3$.
$|4| - |{}^{-}3| = 4 - 3 = 1$
As $|4| > |{}^{-}3|$, the sum will be positive.
$4 + {}^{-}3 = 1$

**Note*: A number plus its opposite will always equal zero.

You can also use a number line to help you find sums when you are adding two numbers that have unlike signs. Use the first addend as your starting point.

- If you are adding a positive number, move to the **right**.
- If you are adding a negative number, move to the **left**.

Examples Add $^{-}1 + 5$.
Start at $^{-}1$ and move to the right 5.

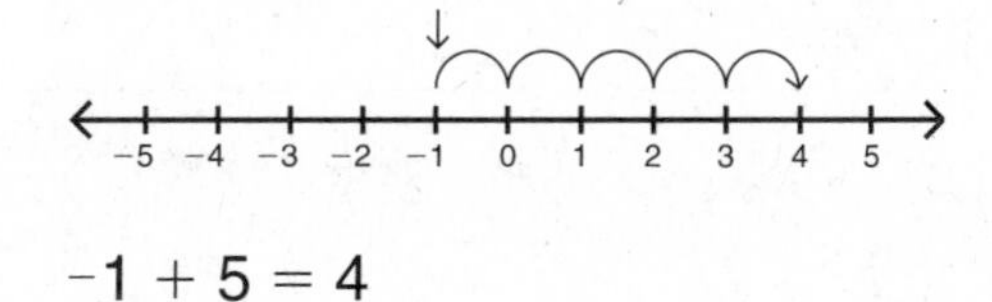

$^{-}1 + 5 = 4$

Add $2 + {}^{-}4$.
Start at 2 and move to the left 4.

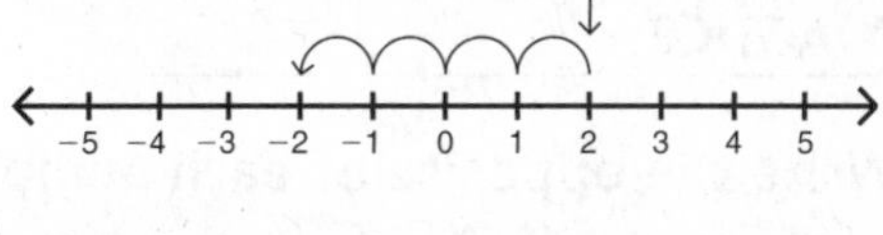

$2 + {}^{-}4 = {}^{-}2$

- When there are three or more addends, add two at a time. The order does not matter.

PRACTICE

Fill in each blank. Use absolute values or a number line to find sums.

1. What is 3 to the left of 0? ________
2. What is 5 to the right of $^{-}1$? ________
3. What is 3 to the left of 2? ________
4. What is 6 to the right of $^{-}5$? ________
5. $^{-}12 + 7 =$ ________
6. $1 + {}^{-}3 =$ ________
7. $^{-}6 + 6 =$ ________
8. $^{-}1 + {}^{-}1 =$ ________
9. $7 + {}^{-}2 =$ ________
10. $^{-}10 + {}^{-}5 =$ ________
11. The temperature was $^{-}2$°F. Then it dropped 2 degrees. What was the temperature then? ________
12. To add $^{-}4$ to 5 on a number line, start at +5 and move 4 spaces to the ________.

Subtracting with Integers

When you subtract a positive number, you take that amount away from the starting amount. The greater the number you subtract, the less you have left. And if you subtract a number greater than the starting amount, you will wind up with a negative number. However, when you subtract a negative number, it is the same as adding the number's opposite. It is possible to end up with a number that is greater than the one you started with. Study the patterns in the following examples.

Examples

$3 - 5 = -2 \rightarrow 3 + -5 = -2$		$-4 - 2 = -6 \rightarrow -4 + -2 = -6$	
$3 - 4 = -1 \rightarrow 3 + -4 = -1$		$-4 - 1 = -5 \rightarrow -4 + -1 = -5$	
$3 - 3 = 0 \rightarrow 3 + -3 = 0$		$-4 - 0 = -4 \rightarrow -4 + 0 = -4$	
$3 - 2 = 1 \rightarrow 3 + -2 = 1$		$-4 - (-1) = -3 \rightarrow -4 + 1 = -3$	
$3 - 1 = 2 \rightarrow 3 + -1 = 2$		$-4 - (-2) = -2 \rightarrow -4 + 2 = -2$	
$3 - 0 = 3 \rightarrow 3 + 0 = 3$		$-4 - (-3) = -1 \rightarrow -4 + 3 = -1$	
$3 - (-1) = 4 \rightarrow 3 + 1 = 4$		$-4 - (-4) = 0 \rightarrow -4 + 4 = 0$	
		$-4 - (-5) = 1 \rightarrow -4 + 5 = 1$	

To subtract with integers:

- Rewrite the subtraction problem as an addition problem in which you **add the opposite** of the number being subtracted. Remember, a number's opposite has the same absolute value, but it has the opposite sign. For example, the opposite of 6 is −6, and the opposite of −8 is 8.
- Then follow the rules for adding integers.

PRACTICE

Write the opposite of each number.

1. −3 ______
2. 17 ______
3. −19 ______
4. −12 ______
5. 11 ______
6. 3 ______

Rewrite each subtraction problem as an addition problem in which you add the opposite of the number being subtracted.

7. $-7 - 4 \rightarrow$ ______
8. $14 - 6 \rightarrow$ ______
9. $-8 - (-4) \rightarrow$ ______
10. $3 - (-9) \rightarrow$ ______
11. $5 - 12 \rightarrow$ ______
12. $13 - (-2) \rightarrow$ ______

Find each difference. Be sure to rewrite each subtraction problem to an addition problem in which you add the opposite.

13. $2 - (-3) =$ ______
14. $-3 - 2 =$ ______
15. $-4 - (-5) =$ ______
16. $2 - 5 =$ ______
17. $3 - (-3) =$ ______
18. $4 - (-3) =$ ______
19. $-7 - (-2) =$ ______
20. $-9 - 1 =$ ______
21. $0 - (-9) =$ ______

Multiplying and Dividing with Integers

Here are the rules for multiplying with positive and negative numbers.

- When you multiply two factors that have the same sign, the product is positive.

positive times positive equals positive	$(+) \times (+) = (+)$	$3 \times 5 = 15$
negative times negative equals positive	$(-) \times (-) = (+)$	$-3 \times -5 = 15$

- When you multiply two factors that have unlike signs, the product is negative.

negative times positive equals negative	$(-) \times (+) = (-)$	$-3 \times 5 = -15$
positive times negative equals negative	$(+) \times (-) = (-)$	$3 \times -5 = -15$

If there are more than two factors, multiply two at a time. Their order does not matter.

The rules for dividing positive and negative numbers are the same as those for multiplying.

- When you divide with two numbers that have the same sign, the quotient is positive.

positive divided by positive equals positive	$(+) \div (+) = (+)$	$15 \div 5 = 3$
negative divided by negative equals positive	$(-) \div (-) = (+)$	$-15 \div -5 = 3$

- When you divide with two numbers that have unlike signs, the quotient is negative.

negative divided by positive equals negative	$(-) \div (+) = (-)$	$-15 \div 5 = -3$
positive divided by negative equals negative	$(+) \div (-) = (-)$	$15 \div -5 = -3$

PRACTICE

Find each product.

1. $2 \times -3 =$ _______
2. $-5 \times -4 =$ _______
3. $-3 \times 4 =$ _______
4. $4 \times -8 =$ _______
5. $6 \times -5 =$ _______
6. $-8 \times -8 =$ _______
7. $-7 \times -5 =$ _______
8. $-2 \times 9 =$ _______
9. $6 \times -4 =$ _______

Find each quotient.

10. $12 \div -3 =$ _______
11. $16 \div -2 =$ _______
12. $-20 \div -2 =$ _______
13. $-14 \div 7 =$ _______
14. $50 \div -25 =$ _______
15. $-32 \div -8 =$ _______
16. $-35 \div -7 =$ _______
17. $42 \div -6 =$ _______
18. $-64 \div 4 =$ _______

Simplify each fraction.

19. $\frac{-2}{-6} =$
20. $\frac{4}{-8} =$
21. $\frac{-6}{9} =$

Using Integers to Solve Word Problems

The problems below involve addition, subtraction, multiplication, or division of signed numbers. Set up and solve each problem. *Hint:* Some of these are two-step problems.

1. Lynn had the following transactions on her credit card: −$62.91, −$9.50, −$3.10, +$3.10. What was the change to her credit card balance?

   ____

2. At 8:00 A.M. the temperature was −4°C. By noon, the temperature had risen 12°, and by 4:00 P.M. it had dropped 7° from the noon temperature. What was the temperature at 4:00 P.M.?

   ____

3. A diver rested at −340 feet. Then she started descending, and 20 minutes later she was at −420 feet. How many feet per minute did she descend?

   ____

4. The elevation of La Paz, Bolivia, is 12,000 feet above sea level. Death Valley, California, has an elevation of −242 feet. How much higher is La Paz than Death Valley?

   ____

5. Find the current balance for this balance sheet.

Previous balance	+$246
Deposit	+$12
Expense	−$90
Current balance	

6. Last month, the books for Joe's restaurant had a balance of $27,000. But this month Joe has lost business because of street repairs being made in front of his restaurant. In the past three weeks, Joe recorded −$800, −$1,200, and −$920. What is his balance now?

   ____

7. Nick has turned on his new freezer. The temperature inside is now 10°F. If the temperature drops 2 degrees each minute, how long will it take the temperature to reach −10°F?

   ____

8. The XYZ Company lost $12,000 in the first quarter of the year. They made a profit of $4,650 in each of the next two quarters. They made $5,025 in the fourth quarter. What were their profits or losses for the year?

   ____

9. A number multiplied −32 is 256. What is the number?

   ____

10. An elevator at the 25th floor went down 16 floors and then up 11 floors. How far from the 25th floor was it?

   ____

Integers Skills Checkup

Circle the letter for the correct answer to each problem.

1. $\frac{-12}{-24}$

   A 2 C $-\frac{1}{2}$
   B $\frac{1}{2}$ D -2
   E None of these

2. $6 + {}^-7 =$

   F 13 H 1
   G -13 J -1
   K None of these

3. $-12 \div {}^-2 =$

   A 6 C 24
   B -6 D -24
   E None of these

4. $|\,5 - 9\,| =$

   F 4 H -4
   G 14 J -14
   K None of these

5. $9 + {}^-3 + {}^-1 =$

   A 5 C 7
   B -5 D -7
   E None of these

6. Which set of numbers is in order from least to greatest?

   A $-12 \quad 0 \quad 5 \quad -30$
   B $-30 \quad -12 \quad 0 \quad 5$
   C $0 \quad 5 \quad -12 \quad -30$
   D $-12 \quad -30 \quad 0 \quad 5$

7. What number is 2 less than -13?

   F -11 H -15
   G 11 J 15

8. Which of the following is greater than $\frac{1}{-3}$?

   A $\frac{1}{-2}$ C 0
   B -1 D $\frac{2}{-3}$

9. Lou had a \$50 credit in his credit card account. Then he spent \$167. Which of these number sentences could you use to find Lou's new balance?

   F $-\$50 + (-\$167) = n$
   G $\$167 + \$50 = n$
   H $\$50 + (-\$167) = n$
   J $-\$50 + \$167 = n$

10. Which of the following expressions has the same value as $-2 - (-6)$?

    A $-2 + 6$ C $-2 - 6$
    B $2 + {}^-6$ D $2 - 6$

Integers Skills Checkup (continued)

11. $10 - 3 =$

F 13 H 7
G −13 J −7
K None of these

12. $-2 - (-9) =$

A 7 C 11
B −7 D −11
E None of these

13. $|3| - |-5| =$

F 2 H −2
G 8 J −8
K None of these

14. $-8 + -3 + 4 =$

A 0 C 8
B −8 D −7
E None of these

15. $\frac{80}{-10} =$

F 8 H $-\frac{1}{8}$
G $\frac{1}{8}$ J −8
K None of these

16. What is the distance between −3 and 7 on a number line?

F 4 H −4
G 10 J 11

17. Which of these decimals is less than −1?

A −0.5 C −0.3
B 0.75 D −2.12

18. Which of the following has the same value as $28 - 52$?

F $-28 + 52$
G $28 - (-52)$
H $-28 + -52$
J $28 + -52$

19. Which of the following does *not* equal −12?

A 12×-1 C -12×1
B -12×-1 D $\frac{-12}{1}$

20. Which product is the greatest?

F -8×-12
G 3×-40
H -200×0
J 20×-4

Ratio, Proportion, and Percent

Writing Ratios

A **ratio** is a comparison of two quantities. A ratio can be written in three ways:

- using the word *to* 1 to 5
- with a colon 1 : 5
- as a fraction $\frac{1}{5}$

Statement	Units Being Compared	Ratio
Ona uses 4 cups of flour **for every** cake she bakes.	cups of flour to cakes	4 to 1 4 : 1 $\frac{4}{1}$
Four **out of** 5 doctors recommend Brand A.	doctors who recommend to doctors who do not	4 to 1 4 : 1 $\frac{4}{1}$
Bob drove 55 miles **per** hour.	miles to hours	55 to 1 55 : 1 $\frac{55}{1}$
You can buy 5 pounds of potatoes **for** $1.49.	pounds to dollars	5 to 1.49 5 : 1.49 $\frac{5}{1.49}$

Order is important. The order of the numbers should follow the order of the words. In fraction form, the first number becomes the numerator and the second number the denominator.

You work with ratios the same way you work with fractions, so writing ratios in fraction form makes them easier to use.

Like fractions, ratios should be simplified to lowest terms. However, if a ratio is an improper fraction, *do not* write it as a mixed number.

PRACTICE

Write each ratio in fraction form using words and numbers. Simplify if possible. The first one has been done for you. (*Remember*, order is important.)

1. Each inch on the map stands for 25 miles. $\frac{\text{inches}}{\text{miles}}$ $\frac{1}{25}$
2. Ricardo earns $2,958 a month. ________
3. Plant 1 evergreen tree for every 3 leafy trees. ________
4. Roses cost $4.00 per dozen. ________
5. It takes me 3 hours to make 24 placemats. ________
6. The plane travels 450 miles per hour. ________
7. Alvaro's car gets 20 miles to the gallon. ________
8. The athletic teams have 12 men for every 8 women. ________
9. They lost 6 games out of 26. ________
10. Max spent $340 on materials and $1,200 on labor. ________

Finding Equal Ratios

You can create equal ratios the same way you create equal fractions. Multiply the numerator and denominator by the same number to create a ratio in higher terms. Divide the numerator and denominator by the same number to create a ratio in simpler terms.

Example

The cost of 12 roses is $10. How many roses will you get for $30? for $50?

- Write a ratio with words and numbers for numerator and denominator. Writing words to represent the units or rates helps you keep track of what the numbers in each ratio represent.
- Multiply numerator and denominator by the same number. You will get 24 roses for $10, 36 roses for $30, and 60 roses for $50.

$$\frac{\text{roses}}{\text{dollars}} \quad \frac{12}{10} = \frac{2 \times 12}{2 \times 10} = \frac{24}{20}$$

$$\frac{12}{10} = \frac{3 \times 12}{3 \times 10} = \frac{36}{30}$$

$$\frac{12}{10} = \frac{5 \times 12}{5 \times 10} = \frac{60}{50}$$

Example

George sells 24 pens for $8. How can he group pens to sell fewer of them at a similar rate?

- Write a ratio with words and numbers for numerator and denominator.
- Divide numerator and denominator by a common factor.

George might sell 12 pens for $4, 6 pens for $2, or 3 pens for $1.

$$\frac{\text{pens}}{\text{dollars}} \quad \frac{24}{8} = \frac{24 \div 2}{8 \div 2} = \frac{12}{4}$$

$$\frac{24}{8} = \frac{24 \div 4}{8 \div 4} = \frac{6}{2}$$

$$\frac{24}{8} = \frac{24 \div 8}{8 \div 8} = \frac{3}{1}$$

PRACTICE

Complete each table to create equal ratios. Multiply both numbers in the first ratio by the same number to create each new ratio.

1. You get 3 cents for every 2 cans you bring to the recycling center.

cents	3	6		12	15		21
cans	2		6			12	

2. A factory builds a certain style of table so that the ratio of the length to the width is 5 to 2.

length	5	10	15				35
width	2		6	8	10	12	

Identifying Proportions

When two ratios are equal, they form a **proportion**. The statement 3:4 = 6:8 is an example of a proportion. The statement $\frac{5}{7} = \frac{10}{14}$ is also a proportion.

One way to tell whether a pair of ratios forms a proportion is to find their cross products. If the cross products are equal, the ratios are equal.

Examples

$\frac{9}{12} \stackrel{?}{=} \frac{15}{20}$ $\qquad$ $\frac{4}{6} \stackrel{?}{=} \frac{24}{30}$

$9 \times 20 = 12 \times 15$ ← cross products → $4 \times 30 \neq 6 \times 24$

$180 = 180$ $\qquad$ $120 \neq 144$

proportion $\qquad$ not a proportion

PRACTICE

Fill in the missing numbers to show the cross multiplication for each pair of ratios. Then find the cross products to determine whether the ratios form a proportion. Circle the number of each pair of ratios that form a proportion.

1. $\frac{4}{7}$ $\frac{16}{28}$

   4 × 28 $\qquad$ 7 × ___

   112 $\qquad$ ______

2. $\frac{8}{13}$ $\frac{24}{39}$

   ___ × 39 $\qquad$ 13 × 24

   ______ $\qquad$ ______

3. $\frac{3}{5}$ $\frac{7}{15}$

   3 × ___ $\qquad$ ___ × 7

   ______ $\qquad$ ______

4. $\frac{36}{64}$ $\frac{9}{4}$

   ___ × 4 $\qquad$ ___ × ___

   ______ $\qquad$ ______

5. $\frac{27}{33}$ $\frac{9}{11}$

   ___ × ___ $\qquad$ 33 × ___

   ______ $\qquad$ ______

6. $\frac{3}{2}$ $\frac{108}{72}$

   ___ × ___ $\qquad$ ___ × ___

   ______ $\qquad$ ______

Use cross products to determine whether the ratios in each pair form a proportion. Circle *Y* if the pair is a proportion. Circle *N* if the pair is not a proportion.

7. $\frac{6}{10} \stackrel{?}{=} \frac{18}{24}$ $\qquad$ Y $\quad$ N

8. $\frac{5}{8} \stackrel{?}{=} \frac{30}{40}$ $\qquad$ Y $\quad$ N

9. $\frac{3}{5} \stackrel{?}{=} \frac{18}{30}$ $\qquad$ Y $\quad$ N

10. $\frac{9}{12} \stackrel{?}{=} \frac{15}{20}$ $\qquad$ Y $\quad$ N

11. $\frac{8}{14} \stackrel{?}{=} \frac{4}{21}$ $\qquad$ Y $\quad$ N

12. $\frac{98}{28} \stackrel{?}{=} \frac{14}{4}$ $\qquad$ Y $\quad$ N

Writing a Proportion

Each of the four numbers in a proportion is called a **term**. For many problems, you will need to write a proportion in which one term is missing, and then find the missing term. It is important to be able to write the proportion correctly. A letter is usually used to take the place of the missing term.

Example If Joel drives 75 miles an hour, how long will it take him to drive 375 miles?

- Write the comparison as a ratio in words. This helps show you where to place the terms in the proportion. $\frac{\text{miles}}{\text{hour}}$
- Write the numbers for the comparison given. $\frac{\text{miles}}{\text{hour}} \quad \frac{75}{1}$
- Write a second ratio to create a proportion. In this problem you are looking for an amount of time. Write a letter in the *hour* position of the second ratio. $\frac{\text{miles}}{\text{hour}} \quad \frac{75}{1} = \frac{375}{n}$

PRACTICE

Each problem below can be solved using a proportion. Write the words for the ratio. Identify what is being solved, then write a proportion for the problem. It is not necessary to answer the question. The first one is done for you.

1. In a diagram, 1 centimeter represents 3 feet. How many centimeters are needed to represent 18 feet?

   Solving for centimeters

   $\frac{\text{cm}}{\text{ft}} \quad \frac{1}{3} = \frac{n}{18}$

2. On sale, 2 yards of cloth cost $8.95. If Mina paid $53.70 for cloth (without tax) how many yards of cloth did she buy?

   Solving for ______________

3. In one day, Lance can build 15 feet of fencing. How long will it take him to build a 60-foot fence?

   Solving for ______________

4. Two cookies have 270 calories. Cara ate 5 cookies. How many calories was that?

   Solving for ______________

5. Jason runs 1 mile in 3 minutes. How many miles will he run in 5 minutes?

   Solving for ______________

6. It takes Ann 15 minutes to read one page. How long will it take her to read 25 pages?

   Solving for ______________

7. Bingo Film Company turns down 25 scripts for every script it accepts. If it accepts 8 scripts, how many scripts will it turn down?

   Solving for ______________

8. For every 3 red triangles Elba uses to create a certain pattern, she uses 9 yellow triangles. If she uses 21 red triangles, how many yellow ones will she use?

   Solving for ______________

Using Cross Products to Solve a Proportion

You can use cross products to find the value of a missing term in a proportion. This is a very useful procedure for solving many problems that involve proportions. First, find the cross products. Then divide both sides of the equation by the number multiplying n to find the value of n.

Example If Joel drives 75 miles an hour, how long will it take him to drive 375 miles?

- Set up a proportion in words and numbers. $\frac{\text{miles}}{\text{hour}} \quad \frac{75}{1} = \frac{375}{n}$
- Find cross products. $75 \times n = 1 \times 375$
  $75n = 375$
- Divide to find the value of n. Divide both sides of the equation by the number n is multiplied by. $\frac{75n}{75} = \frac{375}{75}$
  $n = 5$

Since n represents hours, and $n = 5$, it will take Joel 5 hours to drive 375 miles.

PRACTICE

Find the value of n in each proportion. The first one has been done for you.

1. $\frac{36}{4} = \frac{9}{n}$

   $36 \times n = 4 \times 9$

   $36n = 36$

   $\frac{36n}{36} = \frac{36}{36}$

   $n =$ 1

2. $\frac{8}{14} = \frac{n}{21}$

   ____ $\times 21 =$ ____ $\times n$

   ____ $=$ ____ n

   $n =$ ____

3. $\frac{3}{8} = \frac{15}{n}$

   $3 \times$ ____ $= 8 \times$ ____

   ____ $n =$ ____

   $n =$ ____

4. $\frac{6}{3} = \frac{64}{n}$

5. $\frac{4}{70} = \frac{n}{35}$

6. $\frac{8}{40} = \frac{n}{60}$

7. $\frac{9}{n} = \frac{5}{20}$

8. $\frac{n}{3} = \frac{60}{10}$

9. $\frac{3}{21} = \frac{n}{84}$

10. $\frac{3}{7} = \frac{9}{n}$

11. $\frac{n}{8} = \frac{9}{4}$

12. $\frac{12}{5} = \frac{48}{n}$

13. $\frac{2}{n} = \frac{5}{375}$

14. $\frac{5}{16} = \frac{n}{48}$

15. $\frac{7}{n} = \frac{28}{48}$

Finding a Unit Rate

When the two quantities compared in a ratio have different units of measure, the ratio is called a **rate**. Miles to the gallon, dollars per pound, and words per minute are all examples of ratios that are rates. When the denominator of a rate is 1, the rate is called a **unit rate**.

Examples Sheryl can walk 4 miles in an hour. Her rate of speed is $\frac{\text{4 miles}}{\text{1 hour}}$.

Simon types 95 words per minute. His typing rate is $\frac{\text{95 words}}{\text{1 minute}}$.

One way to find a unit rate is to simplify the ratio so the number in the denominator is 1. To do that, divide both numerator and denominator by the number in the denominator.

Example Larry paid $12 for 3 T-shirts. What was the cost of 1 T-shirt?

$$\frac{\text{dollars}}{\text{T-shirts}} \ \frac{12}{3} = \frac{12 \div 3}{3 \div 3} = \frac{4}{1}$$

It cost $4 for 1 T-shirt.

Another way to find a unit rate is to set up a proportion with the comparison you know as the first ratio. For the second ratio, write 1 in the denominator, and write a letter to represent the amount you are looking for in the numerator. Then use cross products to solve for the missing term.

Example The store is selling 10 cartons of yogurt for $5. What is the cost of 1 carton of yogurt?

- Write the ratio in words. The unit is 1 carton; put *cartons* in the denominator. Then write the ratio in numbers.

$$\frac{\text{dollars}}{\text{cartons}} \ \frac{5}{10}$$

- Write the second ratio to create a proportion. Write 1 in the denominator. Write n in the numerator.

$$\frac{\text{dollars}}{\text{cartons}} \ \frac{5}{10} = \frac{n}{1}$$

- Find cross products and solve for n.

$$5 \times 1 = 10 \times n$$

$$n = \frac{5}{10} = \frac{1}{2} \text{ dollar}$$

One carton of yogurt costs 50 cents.

PRACTICE

Find the unit rate in each problem. The unit rate will have 1 in the denominator.

1. 91 tiles to make 13 hot plates
   How many tiles to make 1 hot plate?

2. 54 packages for 3 cartons
   How many packages for 1 carton?

3. $160 for 4 hours
   How much for 1 hour?

4. $93 for 3 dresses
   How much for 1 dress?

5. 300 miles in 2.5 hours
   How many miles 1 hour?

6. 30 note cards for $6
   How many note cards for $1?

Using Unit Rates to Make Comparisons

You can use unit rates to compare the costs of similar items and find out which costs less per unit. Of course, there may be other factors to consider besides the price per unit. For example, one item might cost a little more per unit but may be better quality. Or, one may cost less per unit but is in a package that contains far more than you need. However, comparing unit rates often helps in making a decision.

Example
You can buy a package of paper towels with 240 sheets for $3.64. Another package of the same brand of towels has 350 sheets and sells for $3.99. Which package is the better buy?

- Set up a proportion for each package to find the cost of one sheet. $\frac{\$}{\text{sheets}}$ $\frac{3.64}{240} = \frac{n}{1}$ $\frac{3.99}{350} = \frac{n}{1}$
- Solve for each missing term. $3.64 = 240n$ $3.99 = 350n$
- Use a calculator to divide. $\frac{3.64}{240} = n$ $\frac{3.99}{350} = n$
- Compare unit prices. $n = 0.015$ $n = 0.011$

The package that costs $3.99 for 350 sheets is the better buy because each sheet costs less.

PRACTICE

Circle the letter in front of the better buy or the comparison that is more favorable. Use a calculator to divide.

1. Which is faster?

   A 132 miles in 2 hours
   B 335 miles in 5 hours

2. Which is better mileage?

   C 356 miles for 12 gallons of gas
   D 616 miles for 22 gallons of gas

3. Which is a better buy?

   F a package of 42 bows for $7.14
   G a package of 120 bows for $25.20

4. Which is the higher salary?

   H $3,400 a month
   J $42,000 a year

5. Which is a better buy?

   A a 24-oz loaf of bread for $3.60
   B a 32-oz loaf of bread for $4.16

6. Which is faster?

   C 504 miles in 8 hours
   D 620 miles in 10 hours

7. Which is a better buy?

   F 25 greeting cards for $8.79
   G 30 greeting cards for $10.29

8. Which is a better buy?

   H 32 oz of applesauce for $2.89
   J 20 oz of applesauce for $2.09

Understanding Percent

A **percent** is a ratio that tells how many out of 100.

Percent means **per hundred**. The symbol for percent is %.

Examples 5% means 5 per hundred, or 5 out of 100, or $\frac{5}{100}$.

37% means 37 per hundred, or 37 out of 100, or $\frac{37}{100}$.

To solve problems, you need to change percents to fractions or decimals. For example, suppose you want to buy something for $1,000 and there is a tax of 5%. Since 5% is 5 hundredths, the tax is equal to $\frac{5}{100}$ or 0.05 of $1,000.

- To write a percent as a decimal, write the percent as a fraction with a denominator of 100. Remove the % sign, and write the number as the numerator of a fraction. Write 100 in the denominator. Once you have a fraction with a denominator of 100, it is easy to write that amount as a decimal.

Examples $25\% = \frac{25}{100} = 0.25$ $100\% = \frac{100}{100} = 1.00$ or 1 $150\% = \frac{150}{100} = 1.50$ or 1.5

Note: 100% is always equal to 1 whole or 1.

- To name a percent as a fraction, remove the % sign and write the number over a denominator of 100. Then reduce the fraction to simplest terms.

Examples $25\% = \frac{25}{100} = \frac{1}{4}$ $40\% = \frac{40}{100} = \frac{2}{5}$ $72\% = \frac{72}{100} = \frac{18}{25}$

PRACTICE

First write each percent as a fraction with a denominator of 100. Then rename that fraction to simplest terms and to a decimal.

Percent	Fraction with Denominator of 100	Fraction in Simplest Terms	Decimal
3%			
5%			
8%			
10%			

Percent	Fraction with Denominator of 100	Fraction in Simplest Terms	Decimal
20%			
50%			
75%			
80%			

Write each percent as a fraction in simplest terms.

1. 6% of the people liked Brand B. What fraction liked Brand B?

2. 75% of the class passed the test. What fraction of the class passed?

3. The coat is marked down 40%. What fraction of the price does that represent?

4. 60% of the people voted for Jones. What fraction voted for Jones?

Writing Decimals as Percents

Since percent means hundredths, a decimal expressed in hundredths can be easily written as a percent. Just replace the word *hundredths* with a percent sign.

Examples 0.05 = 5 hundredths = 5% 0.13 = 13 hundredths = 13%

Notice that the result is like moving the decimal point 2 places to the right and adding the percent symbol.

0.05 ⟶ 0.05. = 5% 0.13 ⟶ 0.13. = 13%

A decimal that is expressed as tenths can be written as an equal amount in hundredths. Write a 0 in the hundredths place.

Examples 0.6 = 0.60 = 60 hundredths = 60% 0.1 = 0.10 = ten hundredths = 10%

0.6 = 0.60 ⟶ 0.60. = 60% 0.1 = 0.10 ⟶ 0.10. = 10%

The shortcut for writing a decimal number as a percent is to move the decimal point 2 places to the right, and then write the percent symbol.

Examples

0.2 ⟶ 0.20 = 20%

0.17 ⟶ 0.17 = 17%

0.304 ⟶ 0.304 = 30.4%

1 ⟶ 1.00 = 100%

15 ⟶ 15.00 = 1,500%

PRACTICE

Write each decimal number as a percent.

1. 0.14 = ________
2. 0.4 = ________
3. 6 = ________
4. 0.013 = ________
5. 0.26 = ________
6. 0.7 = ________
7. 0.1 = ________
8. 0.01 = ________
9. 0.001 = ________
10. 0.9 = ________
11. 0.72 = ________
12. 0.19 = ________

Writing Fractions as Percents

A fraction that has a denominator of 100 can be written as a percent. Just remove the denominator of 100 and write a percent sign after the number.

Examples $\frac{15}{100} = 15\%$ $\frac{3}{100} = 3\%$ $\frac{27}{100} = 27\%$

- If a fraction does not have a denominator of 100, you can rename it. One way to do this is to multiply the numerator and the denominator by the same number. The number you choose to multiply by should be one that produces a product of 100 in the denominator.

Example $\frac{3}{4} = \frac{25 \times 3}{25 \times 4} = \frac{75}{100} = 75\%$

- Another way to rename the fraction is to write a proportion. Write a fraction with a letter in the numerator and 100 in the denominator. Find the cross products and divide to find the missing term. Then replace the letter in your proportion with the number it equals.

Example $\frac{3}{4} = \frac{n}{100}$

$3 \times 100 = 4n$

$\frac{300}{4} = \frac{4n}{4}$

$n = 75$, so $\frac{3}{4} = \frac{75}{100} = 75\%$

- A third way to rename a fraction as a percent is to divide the numerator by the denominator. Then write the resulting decimal as a percent. Remember, percent means hundredths.

Example $\frac{3}{4} = 3 \div 4 = 0.75 = 75\%$

PRACTICE

Write each fraction as a percent.

1. $\frac{1}{10} =$ ________
2. $\frac{3}{5} =$ ________
3. $\frac{1}{2} =$ ________
4. $\frac{3}{20} =$ ________
5. $\frac{1}{4} =$ ________
6. $\frac{1}{8} =$ ________
7. $\frac{3}{8} =$ ________
8. $\frac{3}{15} =$ ________
9. $\frac{8}{10} =$ ________
10. $\frac{5}{8} =$ ________
11. $\frac{10}{20} =$ ________
12. $\frac{1}{5} =$ ________

Finding a Percent of a Number

Remember that in math, the word *of* means multiply. To find a percent of a number, change the percent to a fraction or to a decimal and then multiply.

Examples Find 10% of 300.

Fraction

- Think $10\% \times 300$
- Write 10% as a fraction. $10\% = \frac{1}{10}$
- Multiply. $\frac{1}{10} \times 300 = 30$

10% of $300 = 30$

Decimal

- Think $10\% \times 300$
- Write 10% as a decimal. $10\% = 0.10$
- Multiply. $0.10 \times 300 = 30$

10% of $300 = 30$

PRACTICE

Find each percent. Be sure to label answers for money problems with a dollar sign and to have the correct number of decimal places.

1. 50% of 70 = ________
2. 25% of \$80 = ________
3. 10% of 50 = ________
4. 75% of 40 = ________
5. 15% of 85 = ________
6. 8% of \$600 = ________
7. 100% of \$35 = ________
8. 20% of 45 = ________
9. 5% of 20 = ________
10. 12% of 65 = ________
11. 0.5% of 25 = ________
12. 0.01% of 50 = ________
13. There are 35 people on a committee. To be elected as chairperson, a person must win the votes of 60 percent of the people on the committee. How many votes must a person receive to be elected chairperson?
14. Helen does not want to spend any more than 30% of her income on rent. Her income is \$2,300 a month. What is the highest monthly rent Helen should pay?

Adding or Subtracting a Percent of the Whole

One way to find the cost of something that has been reduced by a certain percent is to find the amount of the reduction. Then subtract that amount from the cost.

Example A blouse that usually sells for $40 is marked down 25%. What is the sale price of the blouse?

- The markdown is 25%. Find 25% of $40. $25\% = \frac{1}{4}$

  $\frac{1}{4} \times \$40 = \$10$

- Subtract the markdown from the regular price. $\$40 - \$10 = \$30$

The sale price is $30.

Here is another way to solve the same problem. The regular cost is 100%, and the blouse is marked down 25%, so the sale price is 75% of the original price. Find 75% of $40.

- The price is 75% of the regular cost. Find 75% of $40. $75\% = 0.75$

The sale price is $30. $0.75 \times \$40 = \$30$

To find the total with a percent added on, first find the amount that has to be added. Then add that amount to the original amount.

Example Bryon's lunch cost $12.00. A tax of 8% was added to that. What was the total cost of the lunch?

- The tax is 8%. Find 8% of $12. $8\% = 0.08$

  $0.08 \times \$12 = \$0.96$

- Add $0.96 tax to the cost of the lunch. $\$12 + \$0.96 = \$12.96$

The total cost of Bryon's lunch was $12.96.

Here is another way to solve the problem. The cost of the lunch is 100%. The tax is 8% of the cost of the lunch; 100% + 8% = 108%. The total cost is 108% of the cost of the lunch.

- Find 108% of $12. $108\% = 1.08$

  $1.08 \times \$12 = \$12.96$

The total cost of Bryon's lunch was $12.96.

PRACTICE

Solve each problem. Label your answers.

1. A $50 dress is marked down 20%. How much does it cost? (*Ignore the tax.*)

2. Hillary gives 10% of her take-home pay to charity. Her yearly take-home pay is $28,000. How much is left after she makes her donations to charity?

3. A magazine costs $3.50 plus 8% tax. What is the final cost to the buyer?

4. In a local election, Cox received 25% more votes than Borland. Borland received 5,420 votes. How many votes did Cox get?

Seeing Percent as a Part to Whole Relationship

Percent is a way of describing part of a whole. Suppose 10 people are asked to name their favorite color, and 7 of them choose blue. All of the people surveyed (10) make up the whole. The number that chose blue (7) represents part of the whole. You can say 7 out of 10, or 70% of the people surveyed, chose blue as their favorite.

Example

$$\frac{\text{part}}{\text{whole}} \quad \frac{\text{people picking blue}}{\text{people surveyed}} = \frac{7}{10} = \frac{70}{100} = 70\%$$

To solve a percent problem, you will either have to find the part, the whole, or the percent. In order to find one, you must know the other two, so it is important to be able to identify which is which.

- Generally, the number right after the word *of*, or right after the multiplication sign, is the whole.
- Generally, the number right before the word *is*, or right before the equal sign, is the part.

Examples

What is 10% of 50?	Whole = 50.	Percent = 10%.	You are looking for the part..
How much is 3% of 12?	Whole = 12.	Percent = 3%.	You are looking for the part.
6 is what percent of 18?	Whole = 18.	Part = 6.	You are looking for the percent.
15 out of 30 is what percent?	Whole = 30.	Part = 15.	You are looking for the percent.
What percent of 80 is 20?	Whole = 80.	Part = 20.	You are looking for the percent.
15 is 10% of what number?	Part = 15.	Percent = 10%.	You are looking for the whole.
25% of what number is 10?	Part = 10.	Percent = 25%.	You are looking for the whole.

PRACTICE

Identify the percent, the whole, and the part in each problem below.

1. 48 is 20% of 240.

   A percent = ________
   B whole = ________
   C part = ________

2. 65% of 200 is 130.

   D percent = ________
   E whole = ________
   F part = ________

3. 7 is 20% of 35.

   A percent = ________
   B whole = ________
   C part = ________

4. 20% of 90 is 18.

   D percent = ________
   E whole = ________
   F part = ________

5. $36 is 20% of $180.

   A percent = ________
   B whole = ________
   C part = ________

6. 40 is 50% of 80.

   D percent = ________
   E whole = ________
   F part = ________

Solving Three Types of Percent Problems

A percent problem involves three numbers: the percent, the whole, and the part. If you know two of these numbers, you can find the third. One way to solve a problem is to write a proportion for a part-to-whole relationship.

- In one ratio, write the whole amount in the denominator and the part in the numerator.
- The second ratio will represent the percent. This ratio will *always* have a denominator of 100.

One term of the proportion will be the one you need to find. Write a letter for that term. Then use cross products to solve for the value of the missing term.

Examples

Finding the Part

600 cars were sold. (whole)
12% of the cars sold were green. (percent)
How many of the cars sold were green? (part)

- Write the relationship as a proportion. Use a letter to represent the term to find. $\frac{\text{part}}{\text{whole}} \rightarrow \frac{\text{green cars}}{\text{cars sold}} \rightarrow \frac{n}{600} = \frac{12}{100}$
- Cross multiply and find the value of n.

$$100 \times n = 12 \times 600$$
$$\frac{100 \times n}{100} = \frac{12 \times 600}{100}$$
$$n = 72$$

72 of the cars sold were green.

Finding the Whole

12% of the cars sold were green. (percent)
72 green cars were sold. (part)
How many cars were sold? (whole)

- Write the relationship as a proportion. Use a letter to represent the term to find. $\frac{\text{part}}{\text{whole}} \rightarrow \frac{\text{green cars}}{\text{cars sold}} \rightarrow \frac{12}{100} = \frac{72}{n}$
- Cross multiply and find the value of n.

$$12 \times n = 72 \times 100$$
$$\frac{12 \times n}{12} = \frac{72 \times 100}{12}$$
$$n = 600$$

600 cars were sold.

Finding the Percent

600 cars were sold. (whole)
72 green cars were sold. (part)
What percent of the cars sold were green? (percent)

- Write the relationship as a proportion. Use a letter to represent the term to find. $\frac{\text{part}}{\text{whole}} \rightarrow \frac{\text{green cars}}{\text{cars sold}} \rightarrow \frac{n}{100} = \frac{72}{600}$
- Cross multiply and find the value of n.

$$600 \times n = 72 \times 100$$
$$\frac{600 \times n}{600} = \frac{72 \times 100}{600}$$
$$n = 12$$

12% of the cars sold were green.

PRACTICE

Circle *part, whole,* or *percent* to tell what you are to find in each problem. Next, write a proportion using the part-to-whole relationship. Have one ratio of the proportion represent percent with 100 as the denominator. Use a letter to represent the term you need to find. Then solve.

1. Randall borrowed $9,000 to buy a new car. So far he has paid $1,800 of his loan. What percent of the loan is paid off?

   Find: part whole percent

2. At Big Company, 245 workers went on strike. That is 35% of the total number of workers at the factory. How many workers are there at Big Company?

   Find: part whole percent

3. Five-eighths of the students in the class passed the science test. What percent of the class passed the test?

   Find: part whole percent

4. Kendra earns $1,200 a month at a part-time job. She puts 20% of her earnings into a savings account. How much does Kendra save each month?

   Find: part whole percent

5. Quinn has 40 hours of free time each week. He spends 6 hours of that free time at a gym. What percent of his free time does Quinn spend at the gym?

   Find: part whole percent

6. In a recent survey 5,240 people said they would like to change careers. That was 25% of the people questioned. How many people were questioned?

   Find: part whole percent

7. Christie uses 30% of her monthly salary for rent. Christie's monthly salary is $2,460. How much does she use for rent?

   Find: part whole percent

8. There were 80 questions on the social studies test. Rebecca got 48 questions right. What percent did she get right?

   Find: part whole percent

9. Julie made a down payment of $45.00 on a new sewing machine. If the down payment is 15% of the selling price, how much does the machine sell for?

   Find: part whole percent

10. The total cost for Theodore's new car was $16,400. If he paid a 20% deposit, how much was his deposit?

    Find: part whole percent

Using Ratio and Percent to Solve Word Problems

Write a proportion for each problem. Use words and numbers. Then solve.

1. The directions for a medicine say to take 3 milliliters for every 25 pounds of body weight. How many milliliters should a person weighing 125 pounds take?

2. Linc bought 40 pounds of ground beef for $120. How much did the ground beef cost per pound?

3. Celine started a diet. If she can lose 3 pounds a week, how many weeks will it take her to lose 51 pounds?

4. Rita spent 50 hours making a set of ceramic tiles. Her client paid her $800. How much did Rita make per hour?

5. A 16-ounce can of Mrs. Smith's soup costs $1.19. A 10-ounce can of Savory brand soup costs 89 cents. Which soup costs less per ounce?

Solve each problem.

6. A $32 can of paint is marked down 20%. What is the reduced price of the paint?

7. The sales tax in Mount Clement is 14%. What is the sales tax on an item that costs $45.00?

8. Jorge bought some vases to sell in his shop. He paid $14 for each vase, and he plans to mark up the price 40% to make a profit. What price will he put on each vase?

9. A teapot that regularly sells for $30 is marked down 15%. How much less will the teapot cost after it is marked down?

10. To get a passing grade on his math test, Ricardo must get 80% of the questions correct. There are 150 questions on the test. How many questions must he get correct?

Ratio, Proportion, and Percent Skills Checkup

Circle the letter for the correct answer to each question. Reduce all fractions to simplest terms.

1. 13% of \$25 =

   A \$325 C \$3.25
   B \$225 D \$2.25
   E None of these

2. 45% of ☐ = 27

   F 60 H 121
   G 30 J 55
   K None of these

3. What percent of 80 is 20?

   A 40% C 20%
   B 25% D 60%
   E None of these

4. 375 is what percent of 500?

   F 25% H 45%
   G 55% J 75%
   K None of these

5. What percent of 80 is 16?

   A 50% C 20%
   B 15% D 16%
   E None of these

6. Which decimal has the same value as 75%?

   F 0.75 H 7.5
   G 0.075 J 7.05
   K None of these

7. A company ordered 452 red pens. They returned 113 of the pens because they were blue instead of red. What percent of the pens were returned?

   A 75% C 33%
   B 25% D 40%

8. Elena bought 2 pounds of sliced ham for \$3.75 per pound. Sales tax is 12%. How much does Elena pay?

   F \$7.50 H \$7.40
   G \$0.90 J \$8.40

9. A news article says that the average American family spends 27% of its income on housing. If the average family income is \$35,000 per year, how much does the average American family spend on housing?

   A \$7,245 per year
   B \$9,450 per year
   C \$2,450 per year
   D \$2,700 per year

10. Audry spends \$600 per month on rent. She makes \$25,000 per year. To the nearest percent, what percent of her income does she spend on rent?

    F 28% H 20%
    G 21% J 29%

11. Which set of numbers is in order from least to greatest?

    A $\frac{3}{5}$, 35%, 3.5, $\frac{5}{3}$, 305%
    B $\frac{3}{5}$, $\frac{5}{3}$, 35%, 3.5, 305%
    C 35%, $\frac{3}{5}$, $\frac{5}{3}$, 3.5, 305%
    D 35%, $\frac{3}{5}$, $\frac{5}{3}$, 305%, 3.5

Ratio, Proportion, and Percent Skills Checkup (continued)

12. What is the value of n in the proportion below?

$$\frac{5}{25} = \frac{8}{n}$$

A 40
B $\frac{40}{25}$
C 200
D 1,000

13. Riley bought 42 ounces of cereal for $12.60. How much did he pay per ounce?

F 30 cents
G 52.9 cents
H 3.2 cents
J 23 cents

14. Which of these statements is true?

A 20% < 15.6%
B 98% < $\frac{2}{3}$
C 0.3 > 25%
D 1 > 112%

15. Company rules state that Jeanette must rest 4 hours for every 6 hours she spends driving. Which of the following proportions could Jeanette use to figure out how many hours of rest she must have during a 32-hour drive?

F $\frac{4}{6} = \frac{32}{n}$
G $\frac{4}{32} = \frac{n}{6}$
H $\frac{n}{4} = \frac{6}{32}$
J $\frac{4}{6} = \frac{n}{32}$

16. In a certain country, the banks exchange 7 curos for every 4 U.S. dollars. How many curos will a traveler get for 140 U.S. dollars?

A 80
B 245
C 980
D 560

17. Which of the following number sentences could you use to find 32% of 67?

F $67 \div 0.32 = n$
G $67 \times n = 32$
H $67 \times 0.32 = n$
J $32 \times 6.7 = n$

18. A couch that normally costs $550 is marked down 20%. Which steps will give you the price of the couch on sale?

A Multiply $550 by $\frac{2}{10}$ (or $\frac{1}{5}$) and then add the product to $550.
B Multiply $550 by 0.2 and then subtract the product from $550.
C Divide $550 by 0.2 and then subtract the quotient from $550.
D Multiply $550 by $\frac{1}{2}$.

19. Which fraction has the same value as 25%?

F $\frac{1}{4}$
G $\frac{1}{25}$
H $\frac{1}{5}$
J $\frac{25}{10}$

20. Which of the following number sentences could you use to find the unknown value in the proportion below?

$$\frac{7}{20} = \frac{n}{45}$$

A $\frac{7 \times 20}{45} = n$
B $7 \times 20 \times 45 = n$
C $7 \times n = 20 \times 45$
D $7 \times 45 = n \times 20$

Algebra

Identifying Number Patterns

Being able to recognize patterns is important to success in mathematics. Being able to extend a pattern can help you solve problems and make predictions. Identifying number patterns is basic to algebra.

PRACTICE

For Questions 1–6, fill in the missing numbers to continue the pattern.

1. Start with 84. Subtract 7.
   84, 77, 70, ______, ______, ______,...
2. Start with 9. Add 6.
   9, 15, 21, ______, ______, ______,...
3. Start with 3. Multiply by 4.
   3, 12, 48, ______, ______, ______,...
4. Start with 100. Divide by 2.
   100, 50, 25, ______, ______, ______,...
5. Start with 4. Multiply by 2, then add 3.
   4, 11, 25, ______, ______, ______,...
6. Start with 7. Subtract 3, then multiply by 2.
   8, 10, 14, ______, ______, ______,...

For Questions 7–14, study the pattern and then complete the rule.

7. 1, 6, 11, 16, 21, 26,...
   Each new number is
   the last number plus ______.
8. 88, 44, 22, 11, 5.5, 2.75,...
   Each new number is
   the last number divided by ______.
9. 86, 79, 72, 65, 58, 51,...
   Each new number is
   the last number minus ______.
10. 1, 3, 9, 27, 81,...
    Each new number is
    the last number times ______.
11. 8, 22, 36, 50, 64,...
    Each new number is
    the last number ______________.
12. 15, 30, 60, 120, 240,...
    Each new number is
    the last number ______________.
13. 500, 100, 20, 4, 0.8, 0.16,...
    Each new number is
    the last number ______________.
14. 7, 22, 37, 52, 67, 82,...
    Each new number is
    the last number ______________.

For Questions 15–20, write the missing number or numbers.

15. 8, 16, 24, ______, 40, ______,...
16. 0, 11, 22, 33, ______, ______, 66,...
17. 1, 3, 2, 4, 3, 5, 4, 6, 5, 7, ______, ______,...
18. 1, 2, 4, 5, 7, 8, ______, ______, 13, 14,...
19. 1, 10, 100, ______, 10,000, ______,...
20. 2, 3.5, 5, ______, 8, ______, 11,...

Reviewing Basic Properties of Numbers

The basic properties for working with numbers apply in algebra. Keep in mind that according to order of operations, when parentheses are used, the operation within the parentheses must be completed first.

Commutative Property of Addition and Multiplication If you change the order of the numbers, the answer will not change.

Examples Addition

$6 + 3 = 3 + 6$

$9 = 9$

Multiplication

$4 \times 8 = 8 \times 4$

$32 = 32$

The commutative property does *not* apply to subtraction or division. If you change the order with these two operations, the answer will not be the same.

Subtraction

$8 - 3 \neq 3 - 8$

$5 \neq -5$

Division

$6 \div 3 \neq 3 \div 6$

$2 \neq \frac{1}{2}$

Associative Property of Addition and Multiplication If you change the way you group the numbers, the answer will not change.

Examples Addition

$(2 + 3) + 4 = 5 + 4 = 9$

and

$2 + (3 + 4) = 2 + 7 = 9$

Multiplication

$2 \times (5 \times 3) = 2 \times 15 = 30$

and

$(2 \times 5) \times 3 = 10 \times 3 = 30$

The associative property does *not* apply to subtraction or division.

Examples Subtraction

$(5 - 2) - 1 = 3 - 1 = 2$

but

$5 - (2 - 1) = 5 - 1 = 4$

$2 \neq 4$

Division

$(27 \div 3) \div 3 = 9 \times 3 = 3$

but

$27 \div (3 \div 3) = 27 \div 1 = 27$

$3 \neq 27$

Distributive Property This property combines both addition and multiplication. It says you can add first and then multiply (see left side of example) *or* you can multiply first and then add (right side of example). Either way, you get the same answer.

Example $5(8 + 2) = (5 \times 8) + (5 \times 2)$

$5 \times 10 = 40 + 10$

$50 = 50$

Properties of Zero

- If you add or subtract 0 from any number, the value of the number is not changed.

  $32 + 0 = 32$ $\quad$ $56 - 0 = 56$

- If you subtract a number from itself, the difference is 0.

  $70 - 70 = 0$

- If you multiply any number by 0, the product is 0.

  $16 \times 0 = 16$ $\qquad\qquad$ $0 \times 5 = 0$

- If you divide 0 by any nonzero number, the quotient is 0.

  $0 \div 8 = 0$

- You cannot divide by zero. No matter what number you divide, when you use multiplication to check the answer, you will always get zero for the product. Therefore, mathematicians say dividing by zero has no meaning.

Properties of One

- If you multiply or divide any number by 1, the value will not be changed.

  $35 \times 1 = 35$ $\qquad\qquad$ $47 \div 1 = 47$

- If you divide any nonzero number by itself, the quotient will be 1.

  $23 \div 23 = 1$

Inverse operations Inverses are opposites. Addition and subtraction are inverse operations. You can use one to undo the other. Multiplication and division are also inverse operations.

Examples	addition/subtraction	$3 + 7 = 10$, so $10 - 7 = 3$
	subtraction/addition	$15 - 7 = 8$, so $8 + 7 = 15$
	multiplication/division	$3 \times 7 = 21$, so $21 \div 7 = 3$
	division/multiplication	$45 \div 5 = 9$, so $9 \times 5 = 45$

PRACTICE

Write +, −, ×, or ÷ to complete each number sentence and make it true.

1. $514 \bigcirc 514 = 1$
2. $74 \bigcirc 74 = 0$
3. $0 \bigcirc 21 = 0$
4. $112 \bigcirc 0 = 112$
5. $817 \bigcirc 0 = 0$
6. $16 \bigcirc 1 = 4 \bigcirc 4$

Write +, −, ×, or ÷ and a number in the $\square$ to make the sentence true.

7. $817 + 35 = 852$, so $852 \square = 817$
8. $718 - 26 = 692$, so $718 \square = 26$

Fill in the numbers to complete each statement.

9. $6 \times (2 + 8) = (6 \times \underline{\qquad}) + (6 \times \underline{\qquad})$

   $6 \times \underline{\qquad} = \underline{\qquad} + \underline{\qquad}$

   $\underline{\qquad} = \underline{\qquad}$

10. $(4 \times \underline{\qquad}) + (4 \times \underline{\qquad}) = 4 \times (3 + 10)$

    $\underline{\qquad} + \underline{\qquad} = 4 \times \underline{\qquad}$

    $\underline{\qquad} = \underline{\qquad}$

11. $8 + 6 = \underline{\qquad} + 8$

    $\underline{\qquad} = \underline{\qquad}$

12. $15 + (\underline{\qquad} + 30) = (15 + 15) + 30$

13. $7 \times 6 + 7 \times (3 + 5) = \underline{\qquad} \times (6 + 8)$

14. $5 \times \underline{\qquad} = 7 \times 5$

Working with Functions

A rule that changes one value to another value is called a **function**. When you apply the rule to a starting number, or **input**, the result is called the **output**. You can examine a set of input/output numbers for a pattern to find out what the rule, or function, is.

Examples

Rule: Subtract 4, then multiply by 2.

Input	25	9	4	50
Output	42	10	0	92

What is the rule?

2 ⟶ 9
8 ⟶ 15
15 ⟶ 22

Rule: Add 7.

PRACTICE

For Numbers 1–4, write the rule for the set.

1. 3 ⟶ 7
   1 ⟶ 5
   5 ⟶ 9

2. 5 ⟶ 11
   1 ⟶ 7
   3 ⟶ 9

3. 35 ⟶ 10
   25 ⟶ 0
   33 ⟶ 8

4. 7 ⟶ 21
   3 ⟶ 9
   5 ⟶ 15

For Numbers 5–8, use the rules to find the missing numbers.

5. Subtract 2 and then multiply by 3.

Input	5	2	4	3	6
Output	9		6		

6. Multiply by 2 and then add 1.

Input	1	2	5	4
Output	3		11	

7. Divide by 2 and then add 5.

Input	12	8	6	10
Output		9	8	

8. Subtract 3 and then multiply by 6.

Input	4	10	8	6
Output	6	42		

Circle the letter of the rule used.

9.

Input	1	2	3	4
Output	6	9	12	15

A Multiply by 5 and then add 1.
B Add 1 and then multiply by 3.
C Add 5.

Using Variables to Write Expressions

In algebra, letters take the place of numbers. These letters are called **unknowns**, or **variables**. A letter can represent a single number, or it can represent many values.

In the table at the right, if we let the letter x represent each *In* number, we can describe each *Out* number as $x - 2$.

In	3	4	5	7
Out	1	2	3	5

A **mathematical expression** is like a phrase in English—it is not a complete thought. An example of an expression would be *three times a number*, or $3 \times n$. A complete thought, or sentence, would be $3 \times n = 15$, *or* $3 \times n > 2$, *or* $3 \times n < 4 + 5$, and so on.

These guidelines can help you write an expression with symbols and variables.

- Use a letter to represent the unknown number.
- Look for signal words that can help you decide which operation to use. (See chart on page 32.)
- Read carefully to decide what is happening. For example, is something being subtracted from the variable, or is the variable being subtracted from something?
- Use operations signs ($+$, $-$, $\times$, $\div$) to show the relationship between the variable and the known amount. Remember, with subtraction and division, the order is important.

Example *a number increased by five* → $n + 5$

The letter n represents *a number*. Since the number is increased by 5, add 5 to it.

Note: When a variable is being multiplied, an expression is usually written without the multiplication sign. So, $3 \times n$ would be $3n$.

PRACTICE

Write each expression using a variable. Show multiplication without using the multiplication sign.

1. a number multiplied by sixteen ________
2. a number increased by twelve ________
3. five times a number ________
4. five more than a number ________
5. the difference between a number and twenty ________
6. one-eighth of a number ________
7. the sum of a number and ten ________
8. half of a number (or a number divided by 2) ________
9. a number plus seventeen ________
10. six less than a number ________
11. a number divided by three (or $\frac{1}{3}$ of a number) ________
12. two less than a number ________

Evaluating Expressions

The value of an expression depends on the value of the variable or variables it contains. To **evaluate** or find the value of an expression, replace the variable with the value assigned to it. Then compute.

Examples

Find the value of the expression $n + 3$.

If $n = 5$, then $n + 3 = 5 + 3 = 8$. If $n = 13$, then $n + 3 = 13 + 3 = 16$.

Find the value of $3y^* - 5$.

If $y = 6$, then $3y - 5 = (3 \times 6) - 5 = 18 - 5 = 13$.

**Note:* A number followed by a letter, such as $3y$, indicates multiplication. 3y means $3 \times y$.

PRACTICE

Find the value of each expression.

For Numbers 1–6, use

$x = 2 \quad n = 3 \quad y = 5 \quad m = 10$

1. $x + y =$ ______
2. $3n + 4 =$ ______
3. $\frac{m}{y} =$ ______
4. $2m =$ ______
5. $6n + y =$ ______
6. $xm - 7 =$ ______

For Numbers 7–12, use

$a = 2 \quad b = 4 \quad c = 6 \quad d = 8$

7. $(b \times d) - c =$ ______
8. $3d - c =$ ______
9. $\frac{(a \times c)}{b} =$ ______
10. $(b + c) \times d =$ ______
11. $\frac{c}{a} + d =$ ______
12. $5c - a =$ ______

Interpreting Expressions and Equations

An **expression** shows the relationship between two terms. An **equation** is a mathematical sentence stating that two expressions are equal.

PRACTICE

Circle the expression that represents each situation.

1. Linda spent d dollars. Then she spent 15 dollars more. How many dollars did Linda spend?

   A $d > d$
   B $d + 15$
   C $\frac{d}{2}$

2. Aisha is 6 inches shorter than her husband. If Aisha is t inches tall, how many inches tall is her husband?

   F $t + 6$
   G $6 - t$
   H $t \times 6$

3. Laurie Ann works 8 hours a day. So far today she has worked x hours. How many hours are left in her workday?

   A $x + 8$
   B $8 - x$
   C $x - 8$

4. The guest speaker is talking to 4 different groups at the school. She will spend m minutes with each group, and she will have to wait 30 minutes before seeing the last group. What is the least number of minutes she will spend at the school?

   F $4m - 30$
   G $4m + 30$
   H $\frac{4m}{30}$

Circle the letter of the correct explanation.

5. Enrique earns \$24,000 per year. In the equation $24{,}000 \div 12 = x$, what does x represent?

   A amount of money he earns per hour
   B amount he earns in 12 years
   C amount he earns each month

6. Regular unleaded gasoline costs \$2.05 per gallon. What does y represent in the equation $\$2.05 \times 10 = y$?

   F cost of 10 gallons of gas
   G amount of gas you can buy for \$10
   H amount of gas used to drive 10 miles

7. Rodney has \$150 to spend on presents. He puts \$50 aside for a gift for his wife. He divides the rest evenly to buy gifts for his three children. In the equation $\$150 - \$50 = n$, what does n represent?

   A number of gifts he will buy
   B amount of money he will spend for each child
   C amount of money he will spend for all three children

8. There are 600 apartments in an apartment complex. There are 75 apartments in each building. In the equation $600 \div 75 = w$, what does w represent?

   F number of people living in the complex
   G number of buildings in the complex
   H number of apartments in the complex

Solving Equations

When you solve an equation, you find the value of the variable. You can solve some equations in your head. For example, with $x + 7 = 10$, you can think, *What number is added to 7 to equal 10?* Since you know your addition facts, you know the answer is 3, so $x = 3$.

To solve a more difficult equation, you need to have a method to follow. One method involves using the inverse operation. Addition and subtraction are inverses of each other. You can use one to undo the other. Multiplication and division are also inverses.

You might think of a scale that is balanced. You want to get the variable alone on one side and a number alone on the other side. To get the variable alone, undo whatever was done to it. To keep the balance, do the same thing on the other side of the equal sign.

Study these examples.

$x + 5 = 47$

5 has been added to x.
Undo addition with subtraction.
Subtract 5.

$x + 5 - 5 = 47 - 5$

To keep the balance, subtract 5 from both sides.

$x = 42$

Example 1

$$\begin{aligned} x + 12 &= 43 \\ -12 &= -12 \\ \hline x + 0 &= 31 \\ x &= 31 \end{aligned}$$

In this equation, 12 has been added to x. Subtract 12.
To keep things balanced, subtract 12 on both sides of the equation.
Adding 0 to a number does not change its value, so $x = 31$.
Check: Replace x with 31 in the equation. $31 + 12 = 43$ is true, so the solution is correct.

Example 2

$$\begin{aligned} n - 7 &= 22 \\ +7 &= +7 \\ \hline n &= 29 \end{aligned}$$

In this equation, 7 has been subtracted from n. Add 7.
Add 7 to both sides of the equation.
Check: $29 - 7 = 22$ is true. The solution is correct.

Example 3

$$\begin{aligned} 35 - n &= 19 \\ +n &= +n \\ \hline 35 &= 19 + n \\ -19 &= -19 \\ \hline 16 &= n \end{aligned}$$

If the variable is being subtracted, add it back before you continue.
In this equation, 19 has been added to n. Subtract 19.
Subtract 19 from both sides of the equation.
Check: $35 - 16 = 19$ is true. The solution is correct.

Example 4

$3y = 108$ In this equation, y has been multiplied by 3. Divide by 3.

$\frac{3y}{3} = \frac{108}{3}$ Divide both sides of the equation by 3.

$y = 36$ Check: $3 \times 36 = 108$ is true. The solution is correct.

Example 5

$\frac{w}{6} = 9$ In this equation, w has been divided by 6. Multiply by 6.

$6 \times \frac{w}{6} = 6 \times 9$ Multiply by 6 on both sides of the equation.

$w = 54$ Check: $54 \div 6 = 9$ is true. The solution is correct.

PRACTICE

Solve each equation. Then check your answer by replacing the variable in the original equation with your solution to see if it makes a true sentence. *Remember:* Use the inverse operation, and do the same thing on both sides of the equation.

1. $x + 9 = 27$ $x =$ ________
2. $y \div 13 = 65$ $y =$ ________
3. $7m = 84$ $m =$ ________
4. $p - 36 = 52$ $p =$ ________
5. $n - 51 = 24$ $n =$ ________
6. $12 + k = 37$ $k =$ ________
7. $\frac{x}{8} = 30$ $x =$ ________
8. $4n = 64$ $n =$ ________
9. $75 = 5w$ $w =$ ________
10. $b \times 8 = 96$ $b =$ ________
11. $24 + g = 49$ $g =$ ________
12. $44 - x = 24$ $x =$ ________

Some equations require two steps. To solve them, first undo the addition or subtraction. Then undo the multiplication or division.

Example $2x + 5 = 15$

$-5 = -5$ First undo the addition. Subtract.

$2x = 10$

$\frac{2x}{2} = \frac{10}{2}$ Next, undo the multiplication. Divide.

$x = 5$ Check: $(2 \times 5) + 5 = 10 + 5 = 15$. The solution is correct.

PRACTICE

Solve each equation. Check your answer.

13. $3x - 2 = 7$ $x =$ ________
14. $4n + 12 = 92$ $n =$ ________
15. $\frac{t}{2} - 8 = 2$ $t =$ ________
16. $5y + 8 = 48$ $y =$ ________
17. $12 + 5n = 62$ $n =$ ________
18. $\frac{b}{5} + 6 = 36$ $b =$ ________

Solving Inequalities

When you solve an equation, you find only one value for the variable. When you solve an inequality, the variable can have many values. To solve an inequality, work the same way that you do to solve an equation.

Examples

If $m + 5 > 8$, find the value of m.

$$\begin{array}{r} m + 5 > 8 \\ -5 = -5 \\ \hline m > 3 \end{array}$$

5 has been added to m. Subtract 5.

Note: Numbers greater than 3 include whole numbers such as 4, 5, 6,... and whole numbers with fractional amounts such as 3.05 or $4\frac{2}{3}$, and so on.

Check your answer. If the solution is correct, replacing the variable with any amount greater than 3 should create a true sentence. Notice that 3 is not an answer, because $3 + 5$ is not greater than 8.

Check: $3.01 + 5 = 8.01$, and $8.01 > 8$

Find n when $25 - n < 12$.

$$\begin{array}{r} 25 - n < 12 \\ +n = +n \\ \hline 25 < 12 + n \\ -12 = -12 \\ \hline 13 < n \end{array}$$

Since n has been subtracted from 25, first add n back.

12 has been added to n. Subtract 12.
Subtract 12 from both sides.

Check your answer. Since n is greater than 13, any number greater than 13 should create a true sentence. We can try 14.

Check: $25 - 14 = 11$, and $11 < 12$

PRACTICE

Find the value of the variable in each inequality.

1. $x - 7 < 25$ ________
2. $m + 14 < 52$ ________
3. $23 + p > 50$ ________
4. $y - 12 < 22$ ________
5. $15 - n > 2$ ________
6. $42 + a < 15$ ________
7. $12 - n > 5$ ________
8. $18 - y < 15$ ________
9. $35 - p > 12$ ________
10. $\frac{x}{5} + 36 > 51$ ________
11. $\frac{y}{3} - 4 < 27$ ________
12. $m + 16 > 32$ ________

Writing Equations and Inequalities

An **equation** is a mathematical sentence that contains an equal sign. It tells you that the expression or amount on one side of the equal sign has the same value as the expression or amount on the other side. To write an equation, translate the words into mathematical symbols with variables. Keep in mind that in math, the word *is* means *equals*.

Examples

5 more than a number n is 13 — $n + 5 = 13$

12 less than a number is 45 — $n - 12 = 45$

An **inequality** is a mathematical sentence that tells you one expression or amount does not have the same value as another expression or amount.

Examples

20 is greater than 5 times a number — $20 > 5n$

three times a number is greater than 7 — $3n > 7$

25 is less than a number plus 4 — $25 < n + 4$

PRACTICE

Circle the equation or inequality that matches the description.

1. 52 is 5 greater than a number x.

   A $52 + 5 = x$ C $52 = x - 5$
   B $52 = 5 - x$ D $52 = x + 5$

2. 4 more than 3 times a number is equal to 34.

   F $4 + 3n = 34$ H $3n - 4 = 34$
   G $34 + 3n = 4$ J $34 + 4 = 3n$

3. 15 times a number p is less than 80.

   A $15 + p < 80$ C $p < 80$
   B $15 \times p < 80$ D $p - 15 < 80$

4. 67 decreased by a number w is -3.

   F $67 + -3 = w$ H $67 + w = -3$
   G $67 - w = -3$ J $w - (-3) = 67$

5. 5 less than a number w equals 12.

   A $5 < w = 12$ C $5 < w - 12$
   B $w - 5 = 12$ D $12 - w < 5$

Write the equation or inequality.

6. 8 is more than a number n ________
7. 15 is equal to 3 times a number n ________
8. 12 is less than 5 times a number y ________
9. a number r divided by 6 equals 18 ________
10. 5 times a number n is equal to 75 ________
11. 14 minus a number n equals 2 ________
12. a number n plus a number p equals 47 ________
13. 2 times a number y is greater than 8 ________
14. a number n divided by 4 is greater than 16 ________

Using Equations to Solve Word Problems

Algebra is a very useful tool for solving word problems. You can write an equation to describe the situation in the problem, and then solve the equation.

Example

Penny wants to go on a trip that will cost $930. She has already saved $150 for the trip. To earn the rest of the money, she took a job where she earns $12.00 an hour. How many hours will Penny have to work to earn enough money to pay for the trip?

- Think: money saved + money earned = $930.
  Money saved = $150. We need to find money earned.
- Write an expression to represent money earned.
  Let n represent number of hours Penny will work. $\$12 \times n$ or $12n$
- Write an equation. $150 + 12n = 930$
- Solve.

$$\begin{aligned} 150 + 12n &= 930 \\ -150 \quad &= -150 \\ \hline 12n &= 780 \end{aligned}$$

$$\frac{12n}{12} = \frac{780}{12}$$

$$n = 65$$

Penny will have to work 65 hours.

Check: $150 + (65 × $12) = $150 + $780 = $930

PRACTICE

Write an equation to describe the situation and then solve.

1. Carl rented a boat at a rate of $14 plus $2.75 per hour. He paid $36. How many hours did Carl have the boat?

2. Simon and Maurice ordered a gift for their parents' anniversary. The gift cost $60. Simon has only $18, so Maurice will have to put up twice as much as he had originally planned. How much money had Maurice originally planned to pay?

3. When a number is multiplied by 7 and 5 is added to the product, the result is 89. What is the original number?

4. Tami spent half of her money on a sweater. She spent $14.60 of what was left having dinner with a friend. If she had $3.90 left, how much money did she start out with?

5. Friendly Auto sold 3 times as many used cars as new cars last year. If Friendly Auto sold a total of 600 cars last year, how many used cars were sold?

6. Big Toy Company had an order for 185 toy trucks. They filled 6 large boxes, each with the same number of trucks, and had one small box with 5 trucks. How many trucks were in each large box?

Algebra Skills Checkup

Circle the letter for the correct answer to each problem.

1. What number goes in the box to make the number sentence true?

   $4 + 5 + \square = 12$

   A 7
   B 3
   C 8
   D 4

2. Tomato soup costs \$0.33 per can. Which of these equations shows how much 4 cans would cost?

   F $\$0.33 \times 4 = x$
   G $\$0.33 \div 4 = x$
   H $4 \div \$0.33 = x$
   J $\$0.33 + 4 = x$

3. Find the rule that will change each "Input" number to the corresponding "Output" number. Then use the rule to find the missing number.

Input	40	26	44	20
Output	20	13	22	

   A 10
   B 0
   C 40
   D 15

4. Which expression shows "4 less than the product of a number and 7"?

   F $4 - (n + 7)$
   G $7n + 4$
   H $\frac{n}{7} - 4$
   J $7n - 4$

5. Which equation has the same solution as $\frac{x}{2} + 9 = 16$?

   A $3x + 12 = 54$
   B $15 + \frac{x}{6} = 18$
   C $4x - 12 = 36$
   D $\frac{8x}{3} = 32$

6. At Joe's, a hot dog special costs \$0.95. In the equation $\$11.44 \div \$0.95 = x$, what will the value of x represent?

   F the cost of 0.95 hot dogs
   G the money you would have left after buying a hot dog
   H the number of hot dogs you can buy for \$11.44
   J the price of 5 hot dogs

7. If you start with 3, then add 3 to that number, and repeat adding 3 to each new number, which number will *never* be in the pattern?

   A 9
   B 18
   C 15
   D 22

8. Marcus bought a television that cost \$362. He paid \$75 down and plans to pay the balance in 7 equal payments. How much will each payment be?

   F \$48
   G \$41
   H \$37.50
   J \$23.75

9. Elliot has 15 boxes of candy. He gives 3 to his mother and splits the remaining boxes with his best friend. Which of these equations shows how to find out how many boxes Elliot has left?

   A $\frac{15}{2} - 3 = x$
   B $15 - \frac{3}{2} = x$
   C $\frac{15}{2} + 3 = x$
   D $\frac{15 - 3}{2} = x$

10. Which of the following is *not* a solution for $3x - 10 < 14$?

    F -5
    G 12
    H 4.5
    J 0

Algebra Skills Checkup (continued)

11. Which rule describes the pattern?

Input	0	5	10	15
Output	2	17	32	47

A add 2, multiply by 3
B divide by 5, add 12
C subtract 8, multiply by 4
D multiply by 3, add 2

12. City workers picked up yard waste on April 6, April 15, April 24, and May 3. If this pattern continues, what will be the next date they pick up yard waste?

F May 6 H May 12
G May 14 J May 17

13. Gary worked 55 hours. He earned \$14 per hour, and he got a \$100 bonus. Which of these shows how to find his total earnings?

A $(55 \times \$14) + \$100 = n$
B $55 \times (\$14 + \$100) = n$
C $55 + \$14 + \$100 = n$
D $(55 \times \$14) + (55 \times \$100) = n$

14. Which number belongs in the box?

$$7 \times \square = 0$$

F 7 H 1
G $\frac{1}{7}$ J 0

15. What number is missing from the table?
Rule: Divide by 2 and then add 1.

Input	10	12	16	18
Output	6	7	9	

A 9 C 12
B 10 D 19

16. In which equation is x equal to 12?

F $x \div 2 = 24$ H $x - 5 = 17$
G $x \times 3 = 4$ J $x + 6 = 18$

17. In which equation is x equal to 8?

A $8 - 1 = x$ C $8 \div 1 = x$
B $8 \times 0 = x$ D $8 + 1 = x$

Use the information below to answer Questions 18 and 19.

There are 3 women on Tanika's work crew. Yesterday they worked 6 hours and they assembled 500 CD players.

18. Tanika used this equation to figure out the rate to assemble CDs.

$$500 \div 6 = n$$

What does the n represent in the equation?

F CD players made per worker
G CD players made per hour
H minutes spent on each CD player
J CD players made each minute

19. Each woman on the crew earns \$8.50 per hour. Which of the following equations could you use to figure out the crew's total earnings yesterday?

A $\$8.50 \times 3 = n$
B $\$8.50 \times 6 = n$
C $\$8.50 \times 3 + 6 = n$
D $\$8.50 \times 3 \times 6 = n$

20. What number is missing from the number pattern in the box?

0, 7, 3, 10, 6, 13, ____, 16

F 11 H 9
G 14 J 8

Geometry

Reviewing Geometry Terms

Geometry is the branch of mathematics that deals with lines, points, curves, angles, surfaces, and solids, and with relationships among these things. The chart contains some basic terms used in geometry.

Term	Definition	Symbol
point	a location on an object or a position in space	a point labeled with a capital letter Y• X• Z•
line	a connected set of points extending forever in both directions	P Q $\overleftrightarrow{PQ}$ (or $\overleftrightarrow{QP}$) is a line. A line is named by two points on the line.
line segment	two points (endpoints) and the straight path between them	S T $\overline{ST}$ (or $\overline{TS}$) is a line segment. A line segment is named by its endpoints.
ray	part of a line that extends in one direction	$\overrightarrow{WX}$ is a ray. The endpoint is always named first.
angle	a figure formed by two rays or line segments that meet at a common endpoint	A B C The angle symbol can be followed by three letters. The letter at the common endpoint should be in the middle. $\angle ABC$ or $\angle B$

PRACTICE

Circle the term that identifies each of the following.

1. $\overline{GH}$
   A ray
   B line
   C line segment
   D angle

2. $\overline{MN}$
   F ray
   G line
   H line segment
   J angle

3. $\overleftrightarrow{JK}$
   A ray
   B line
   C line segment
   D angle

For Questions 4–7, draw the figure named.

4. Point C

5. $\overline{AB}$

6. $\overleftrightarrow{ST}$

7. $\angle MOP$

Recognizing Types of Angles

The point where the two rays of an angle meet is called the **vertex**. The distance between the rays is the size or measure of the angle. You can find a more precise measurement using a tool called a protractor, but there are general terms to describe the measure of an angle.

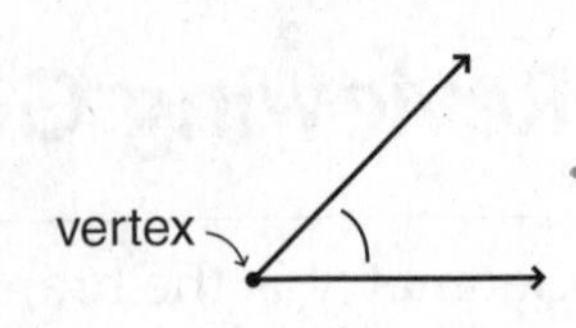

- Angle measurements are related to degrees of a circle. A full circle has 360º. Half of a circle is 180º or 360 ÷ 2. A 180º angle forms a **straight angle** or straight line.

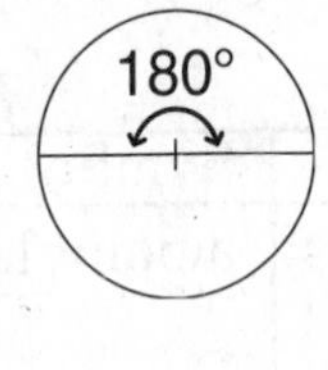

- One-fourth of a circle is 90º (360 ÷ 4). A 90º angle is called a **right angle**. A right angle symbol looks like a square corner.

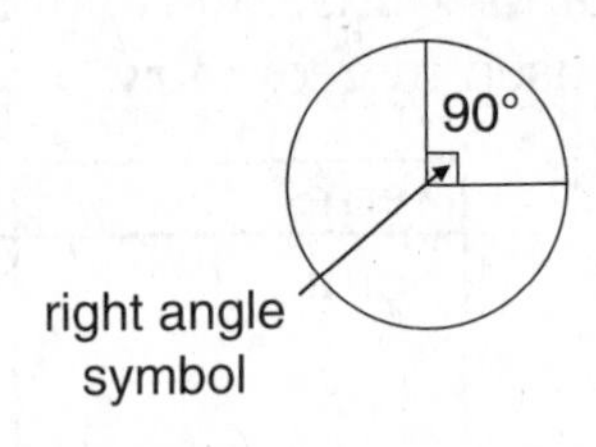

- **Acute angles** are angles that are less than 90º.

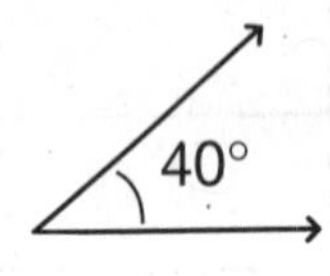

An acute angle is less than 90°.

- **Obtuse angles** are greater than 90º but less than 180º.

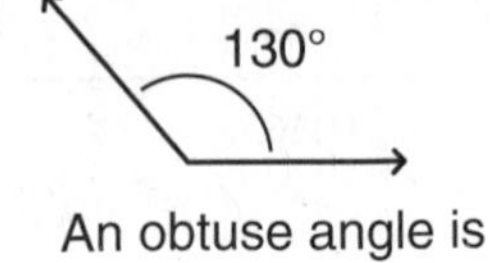

An obtuse angle is greater than 90°.

PRACTICE

Identify each angle as *acute, right,* or *obtuse*.

1. 1. ____________

2. ____________

3. ____________

4. 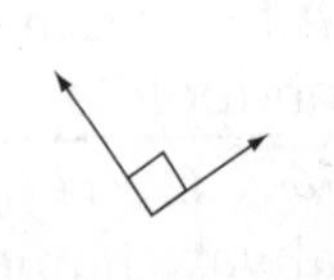 ____________

5. 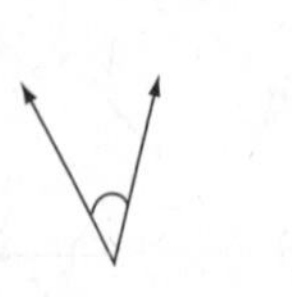 ____________

6. 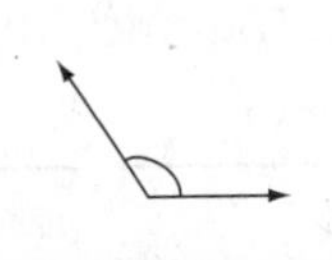 ____________

7. 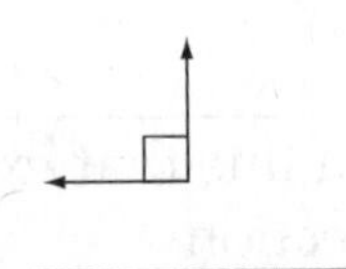 ____________

8. ____________

9. ____________

10. 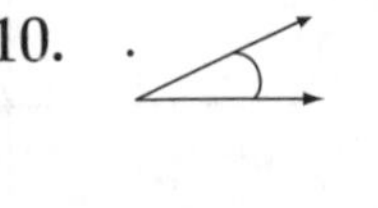 ____________

11.  ____________

12. 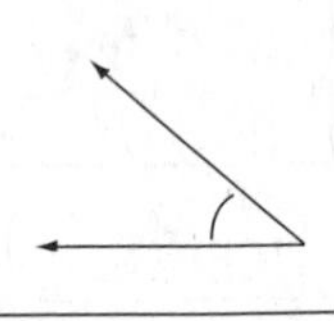 ____________

Use this diagram for Numbers 13 and 14.

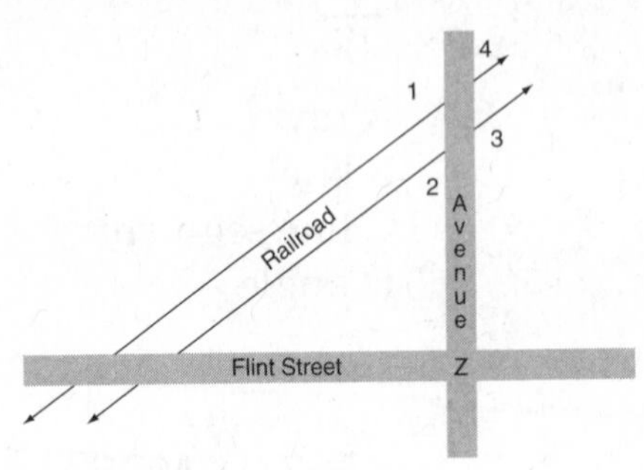

13. The 4 angles formed at the intersection of Flint Street and Avenue Z are

    A right angles **C** obtuse angles
    B acute angles **D** straight angles

14. The angles formed by the intersection of the railroad and Avenue Z are numbered. Which angles are acute?

    F 1 and 2 **H** 1 and 4
    G 2 and 3 **J** 2 and 4

Recognizing Relationships of Lines

Two lines that cross or that will cross are called **intersecting lines**.

Intersecting Lines

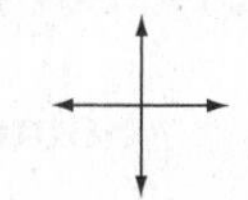 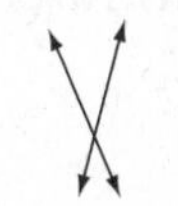 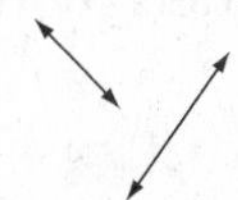

Two lines that are always the same distance apart are called **parallel lines**. The symbol for parallel is ||.

Parallel Lines

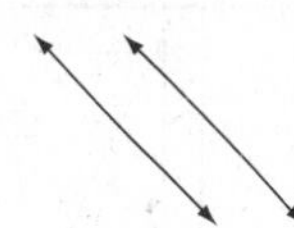

Two lines that form right angles when they meet are called **perpendicular lines**. The symbol for perpendicular is ⊥.

Perpendicular Lines

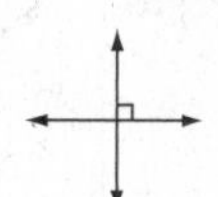 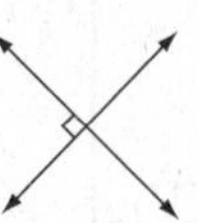

PRACTICE

Write *P* if the lines are parallel. Write *I* if the lines are intersecting.

1. 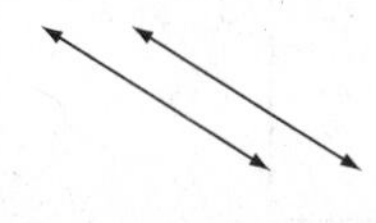 ________

2. 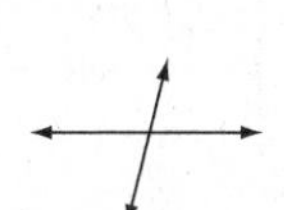 ________

3. 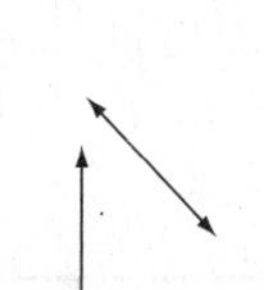 ________

4. 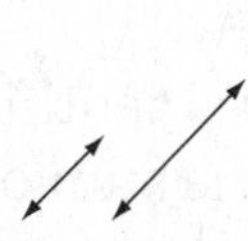 ________

Write *yes* if the lines are perpendicular. Write *no* if they are not perpendicular.

5. 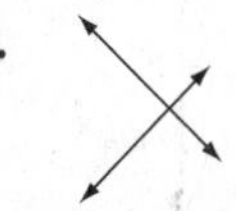 ________

6. 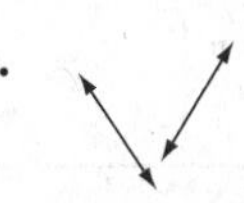 ________

7. 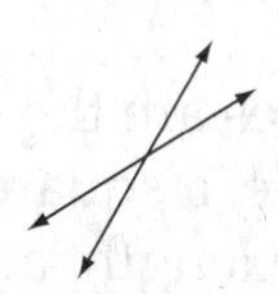 ________

8. 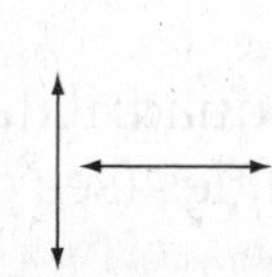 ________

Use this diagram to answer Numbers 9–11.

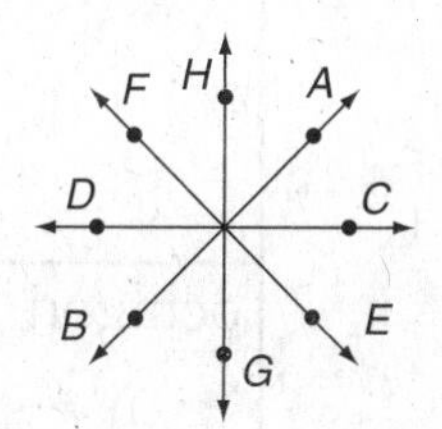

9. Which line or lines are perpendicular to $\overleftrightarrow{AB}$?

____

10. Is $\overleftrightarrow{FE}$ perpendicular to $\overleftrightarrow{DC}$?

____

11. Are there any parallel lines in this figure? If so, name the lines.

____

Identifying Polygons

A **polygon** is a two-dimensional shape that is closed and has straight sides. Polygons are classified by the number of sides they have.

Some Common Polygons

Name	Examples	Number of Sides	Number of Angles
triangle		3	3
quadrilateral		4	4
pentagon		5	5
hexagon		6	6
octagon		8	8

To name a polygon, list the letters of the vertices in order. For example, you can name the figure at the right *ABCE, BCEA,* or *CBAE,* but ***not*** *ACBE.*

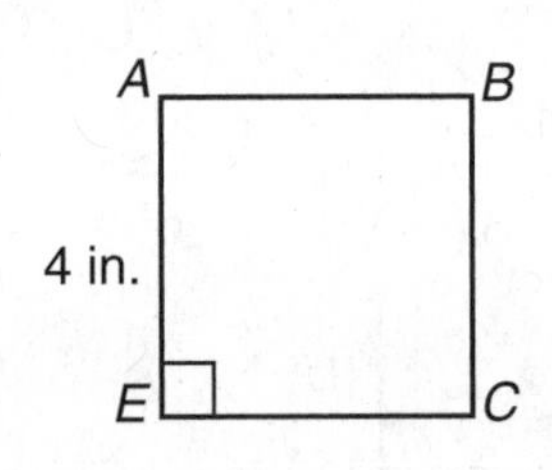

A **regular polygon** is one in which all of the sides are the same length, and the angles are all the same size. For example, a square is a regular quadrilateral; it has four sides that are the same length, and each of its four angles measures 90°.

Several special polygons belong to the family of quadrilaterals. These figures are identified by the relationship of their sides, and by their angles (see chart on next page). The sum of the four angles in every quadrilateral is 360°.

Quadrilaterals	Definition	Examples
parallelogram	a quadrilateral with opposite sides parallel opposite sides and opposite angles are equal	
rhombus	a parallelogram with four equal sides	
rectangle	a parallelogram with four right angles	
square	a rectangle with four equal sides	
trapezoid	a quadrilateral with exactly one pair of parallel sides	

PRACTICE

There are three terms next to each figure. One of the three terms *does not* describe the figure. Circle the letter for the term that *does not* apply.

1. **A** quadrilateral
   B parallelogram
   C square

2. **F** parallelogram
   G regular
   H hexagon

3. **A** quadrilateral
   B parallelogram
   C rectangle

4. **F** trapezoid
   G parallelogram
   H rectangle

5. **A** pentagon
   B regular
   C octagon

6. **F** quadrilateral
   G parallelogram
   H trapezoid

7. **A** polygon
   B regular
   C triangle

8. **F** triangle
   G regular
   H parallelogram

9. **A** polygon
   B regular
   C quadrilateral

10. **F** pentagon
    G hexagon
    H polygon

Recognizing Types of Triangles

Triangles are a particularly interesting type of figure. The sum of any two sides of a triangle must be greater than the third side.

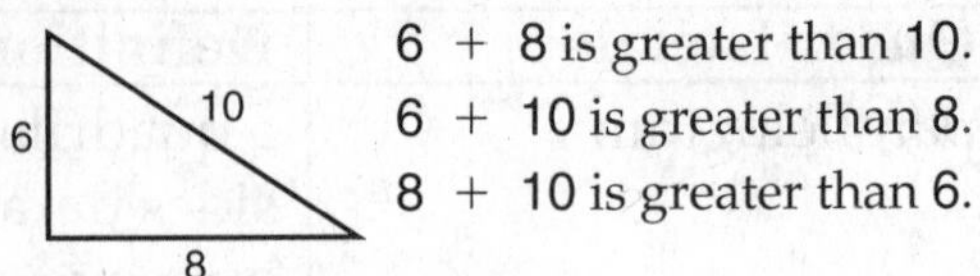

6 + 8 is greater than 10.
6 + 10 is greater than 8.
8 + 10 is greater than 6.

The sum of the three angles of any triangle is always 180°.

180°

Triangles can be classified by the lengths of their sides
- An **equilateral triangle** has three equal sides (and three equal angles).
- An **isosceles triangle** has two equal sides (and two equal angles).
- A **scalene triangle** has no equal sides (and no equal angles).

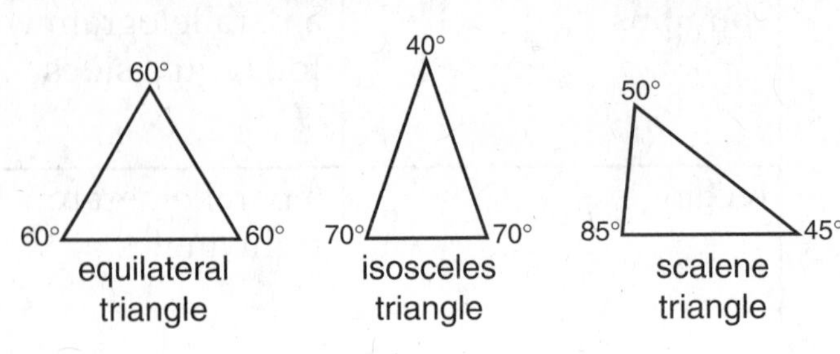

Triangles can also be classified by their angles:
- An **acute triangle** has three acute angles.
- A **right triangle** has exactly one right angle.
- An **obtuse triangle** has exactly one obtuse angle.

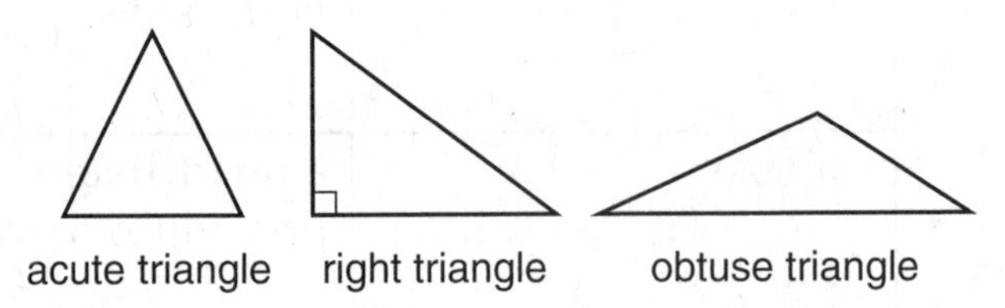

PRACTICE

For each triangle, give the measure of $\angle A$. Then identify the type triangle by its sides and by its angles.

1.
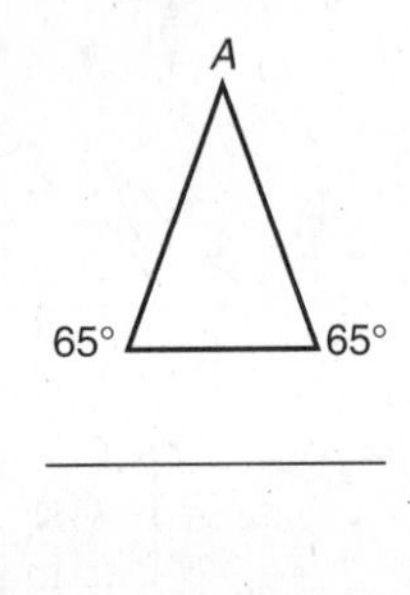

______
______

2.
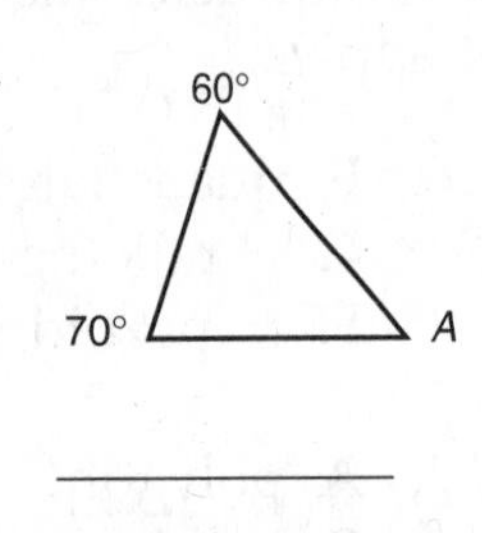

______
______

3.
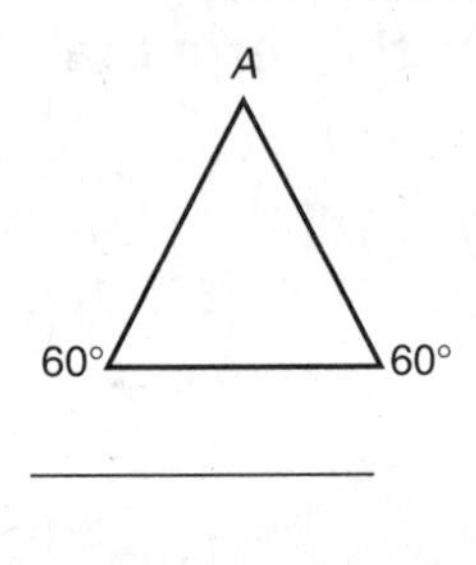

______
______

4.
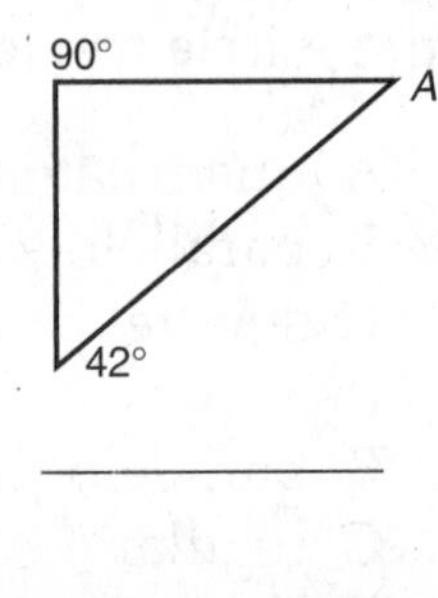

______
______

Below, sides of a triangle that have the same length are marked with a symbol. Study each triangle. Then circle the letter for the term that *does not* apply to the triangle.

5. **A** right
   B acute
   C equilateral

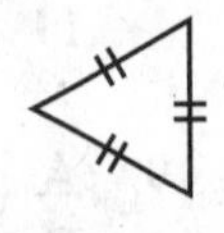

6. **F** right
   G scalene
   H obtuse

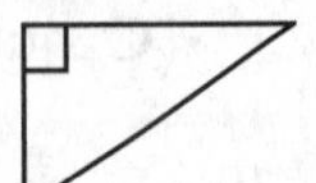

7. **A** obtuse
   B isosceles
   C scalene

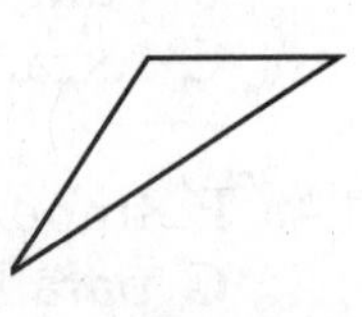

8. **F** acute
   G isosceles
   H equilateral

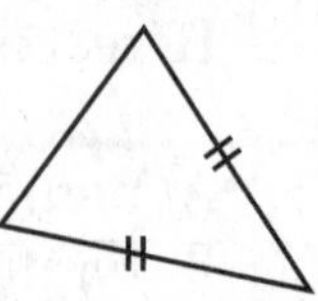

Drawing Diagonals in a Polygon

A **diagonal** is a line segment drawn inside a polygon that connects two vertices of the figure. A side of the figure cannot be a diagonal.

diagonal

You can draw diagonals inside polygons with more than three sides to create triangles. Draw as many diagonals as possible from one vertex. In a quadrilateral, for example, only one diagonal can be drawn from a single vertex.

When you know how many triangles a polygon contains, you can find the total number of degrees in the angles of the figure. A quadrilateral contains two triangles. You know there are 180° in a triangle, so the sum of degrees in the quadrilateral is $2 \times 180°$, or 360°.

PRACTICE

Complete the chart.

Figure	Number of Sides	Diagonals from One Vertex	Number of Triangles	Total Number of Degrees
triangle	3	0	1	180°
quadrilateral	4	1	2	$2 \times 180° = 360°$
pentagon	5	2	3	$3 \times 180° = 540°$
hexagon	6			
octagon				

In a **regular polygon,** the angles are congruent. To find the number of degrees in one angle of a regular polygon, divide the total number of degrees by the number of angles. For example, to find the number of degrees in each angle of square, divide 360° by 4. Each angle is 90°.

PRACTICE

Find the number of degrees in each angle of the following regular polygons.

1. pentagon ________
2. hexagon ________
3. octagon ________

Identifying Parts of a Circle

A **circle** is a closed 2-dimensional shape. All of the points of a circle are the same distance from its center. The following terms describe parts of a circle.

- **Circumference** (*C*) is the distance around the edge of the circle.
- **Radius** (*r*) is the distance from the center of the circle to its circumference.
- A **chord** is a straight line from one point on the circumference to another.
- A **diameter** (*d*) is a chord that passes through the center of the circle. The diameter is the chord in the circle with the greatest length.

circumference

center

radii: $\overline{AC}$, $\overline{BC}$, $\overline{CD}$

diameter: $\overline{BD}$

chords: $\overline{MN}$, $\overline{BD}$

The diameter of a circle is twice the length of its radius. $d = 2 \times r$

The radius of a circle is equal to one-half of its diameter. $r = \frac{d}{2}$

PRACTICE

Complete the questions.

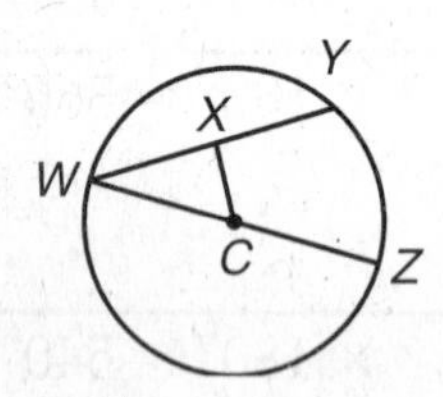

1. Which segment is a radius of the circle above?

   A $\overline{CZ}$
   B $\overline{WY}$
   C $\overline{XY}$
   D $\overline{CX}$

2. Which segment names a chord?

   F $\overline{CZ}$
   G $\overline{WY}$
   H $\overline{WX}$
   J $\overline{CX}$

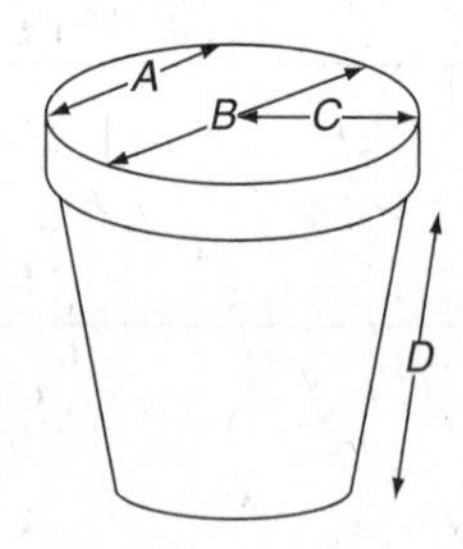

3. Jessica needs a flowerpot that is 6 inches in diameter. Which line shows what she should measure?

   ____

4. Jessica wants to place trim around the outside of the flowerpot. Which term names the part of the flowerpot she will trim?

   ____

Write the diameter and the radius of each circle below. If you do not have enough information to do that, write "cannot tell."

5.

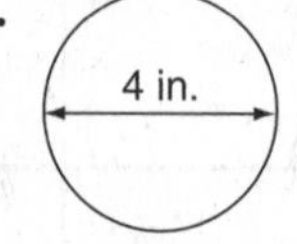

radius: __________

diameter: __________

6.

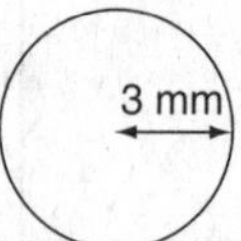

radius: __________

diameter: __________

7.

3 cm

radius: __________

diameter: __________

Recognizing Similarity and Congruence

Two figures that are exactly the same shape, but not necessarily the same size, are **similar**. **Congruent** figures are exactly the same shape *and* the same size. Figures do not have to be facing the same way to be similar or congruent.

PRACTICE

For Numbers 1–5, circle the letter of the figure described.

1. the figure that is congruent to the figure in the first box

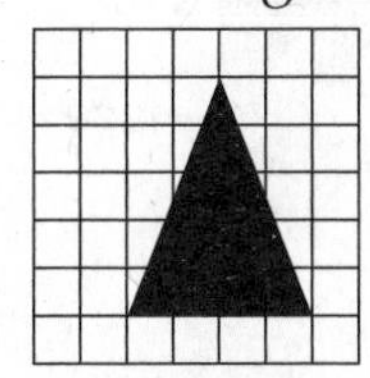

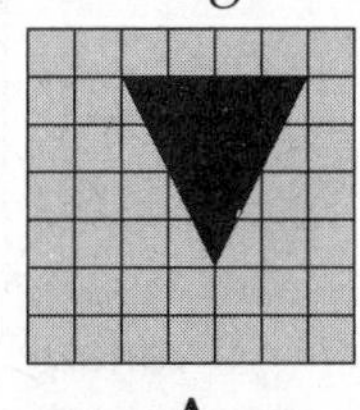

A

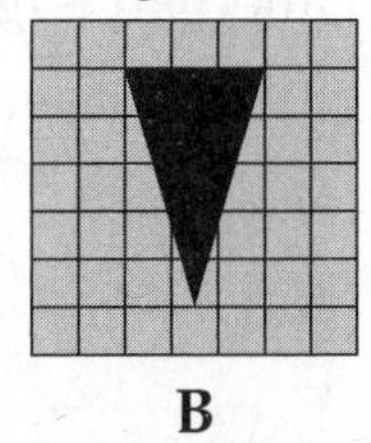

B

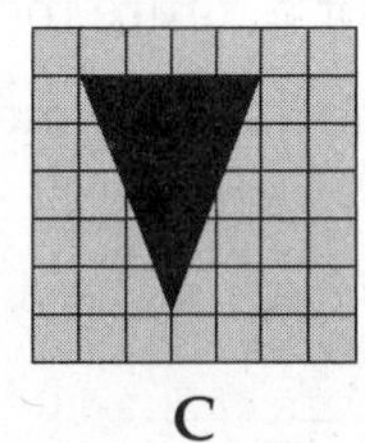

C

2. the figure that is similar to, but not congruent to, the figure in the first box

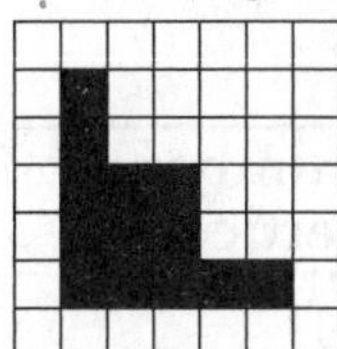

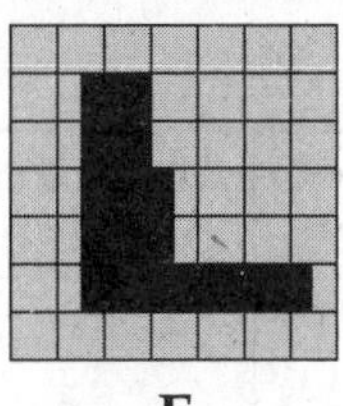

F

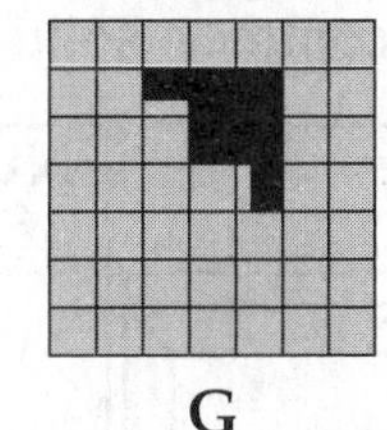

G

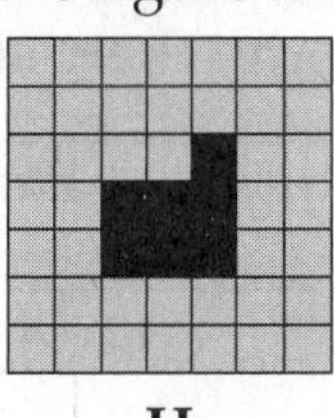

H

3. the figure that is both similar and congruent to the figure in the first box

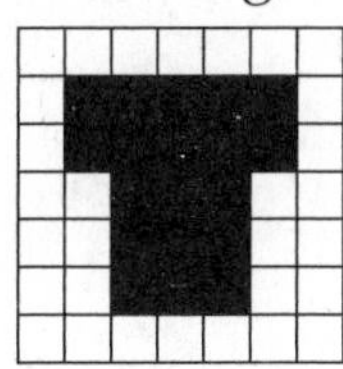

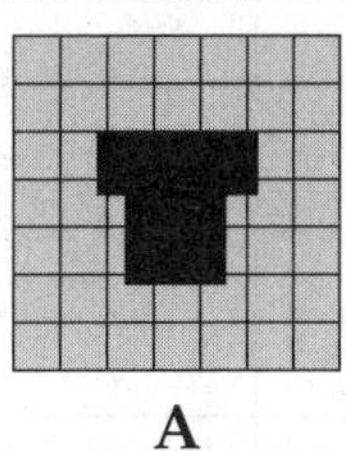

A

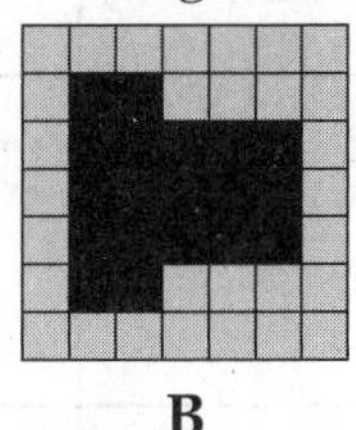

B

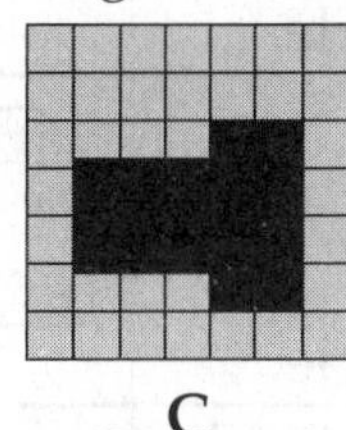

C

4. the figure that is divided into two congruent halves

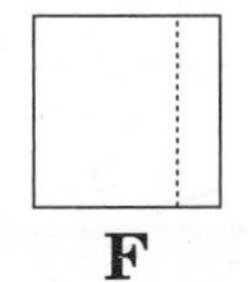

F

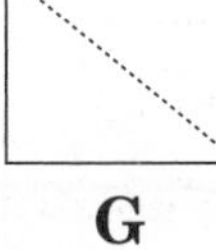

G

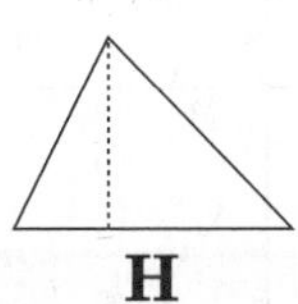

H

5. two congruent sections of this figure

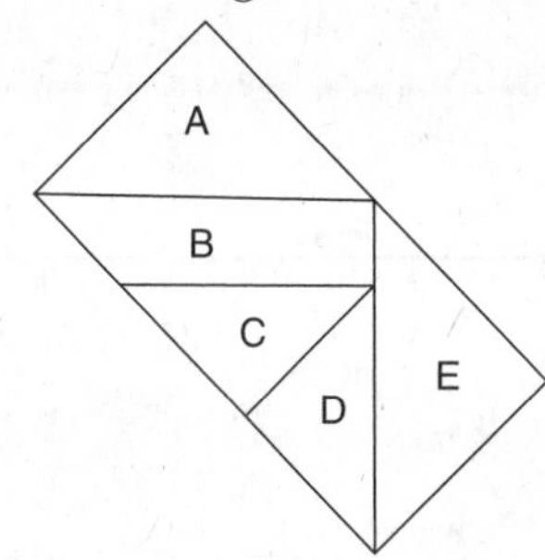

6. In the gray box, draw a figure that is congruent to the dark figure in the white box. Use the grid in the box to make sure your drawing is the same shape and the same size.

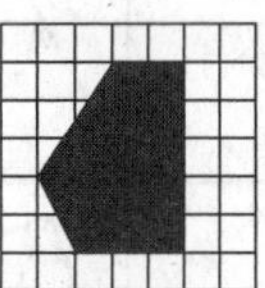

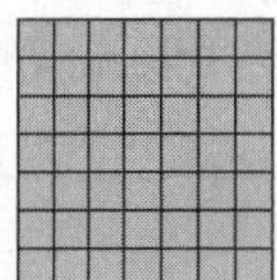

7. In the gray box, draw a figure that is similar to the dark figure, but whose sides are half the length of the dark figure's sides.

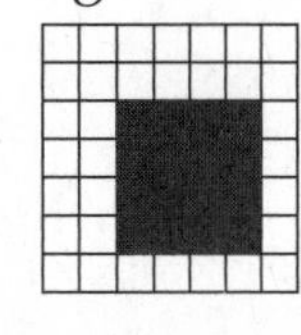

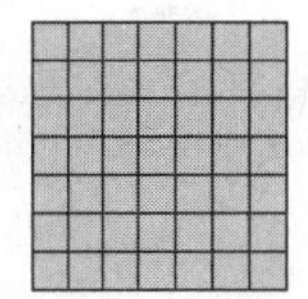

Working with Three-Dimensional Figures

Three-dimensional figures, or **solid figures**, have three dimensions; length, height, and width.

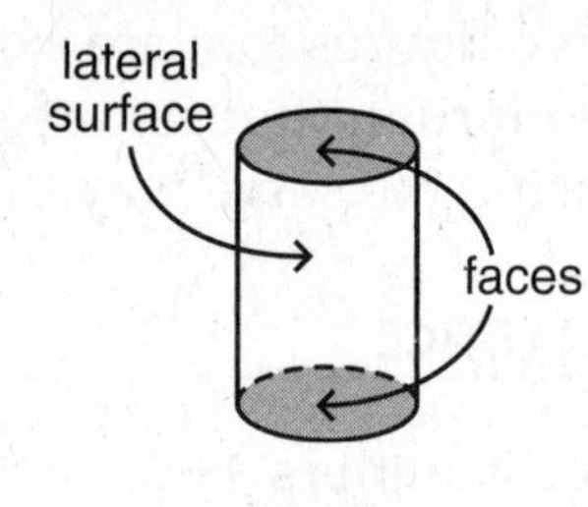

- A flat surface on the outside of three-dimensional figure is called a **face**. Faces have shapes of two-dimensional figures. In a cylinder, the two circular faces are parallel. The curved surface that connects them is called a **lateral surface**.
- The line where two faces of a figure meet is called an **edge**.
- The point where two or more edges meet is called a **vertex**.

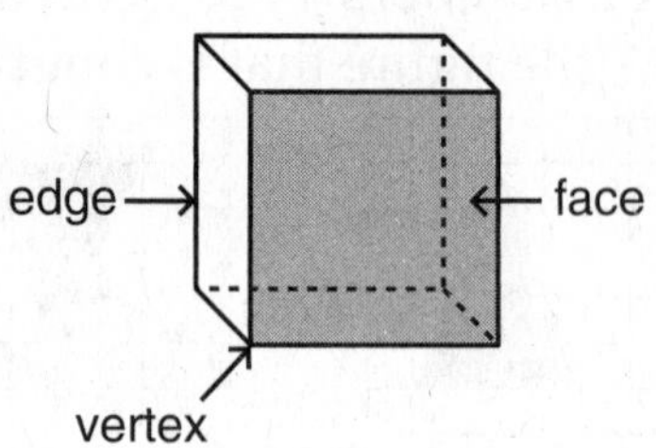

PRACTICE

Complete the information in the chart about common solid figures.

Name of Figure and Description	Example	Number of Faces	Number of Edges	Number of Vertices
rectangular prism (box) All faces are rectangles. Opposite faces are parallel. All corners form a right angle.				
cube All faces are square. Opposite faces are parallel. All corners form a right angle.				
square pyramid Base is a square. All other faces are triangles.				
triangular pyramid Base is a triangle. All other faces are triangles.				
triangular prism One pair of triangular faces are parallel. All other faces are rectangular.				

Some three-dimensional figures have curved surfaces.

- A **cylinder** has two circular faces that are parallel. The surface between the circular faces is rectangular.

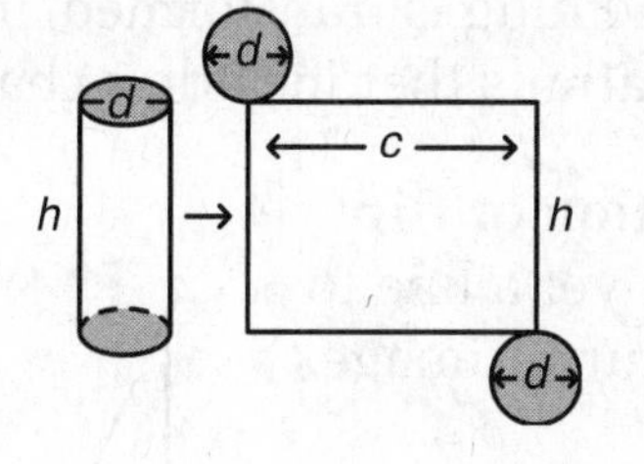

- A **cone** has one circular face. The curved surface ends in a point or vertex.

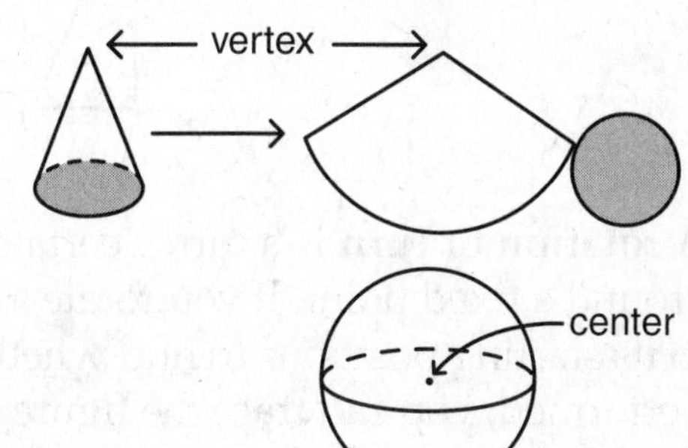

- A **sphere** has a curved surface that looks like a ball or globe. All of the points on the sphere are the same distance from the center of the sphere.

center

PRACTICE

Name the solid that the object named most closely resembles.

1. 

______

2. 

______

3. 

______

4. 

______

5. 

______

6. 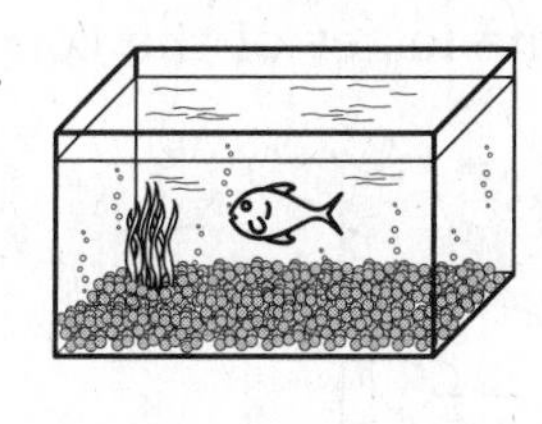

______

Name the solid that would be made by folding each pattern.

7. 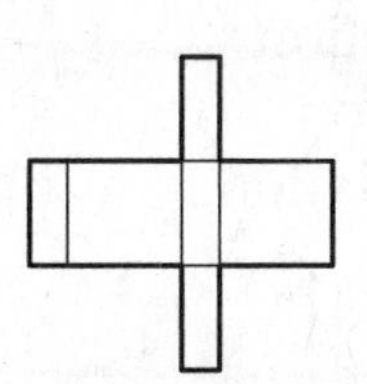

______

8. 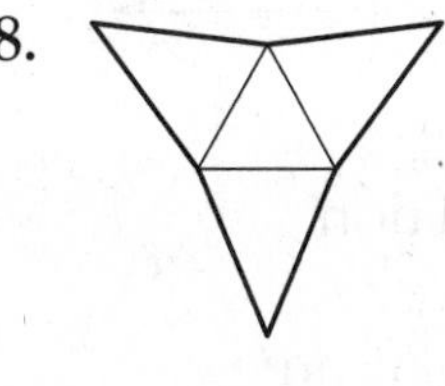

______

9. 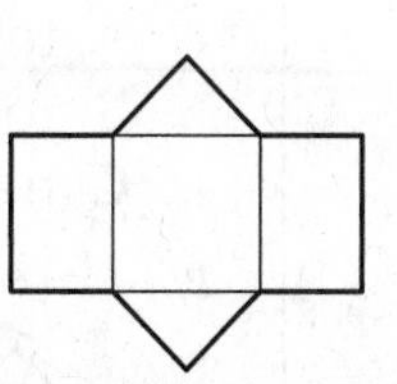

______

10. 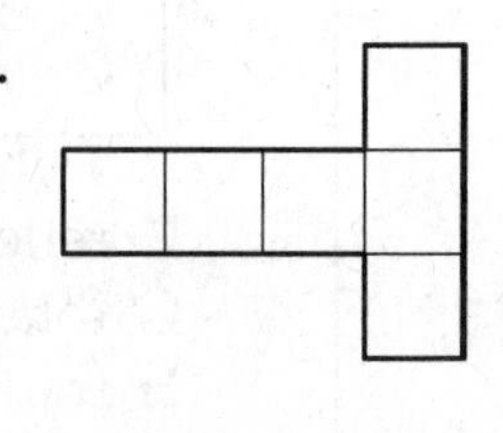

______

Distinguishing Among Transformations

When something is transformed, it is changed in some way. In geometry, there are three **transformations** that include a change in position but not a change in shape or size.

- **A reflection or flip** is a flip over a line to form a mirror image.

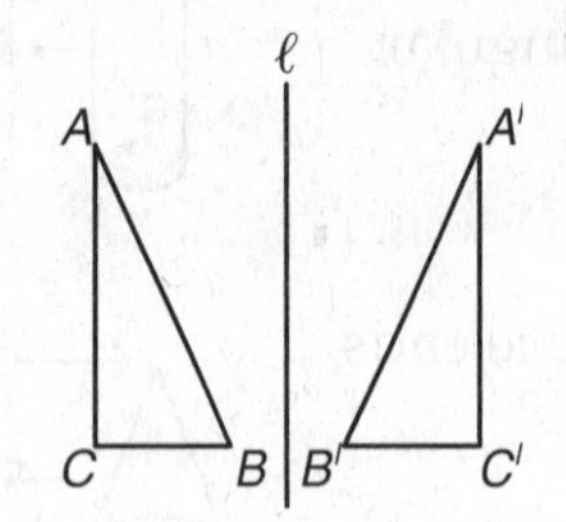

- **A translation or slide** is a move to another position without rotating or reflecting the figure.

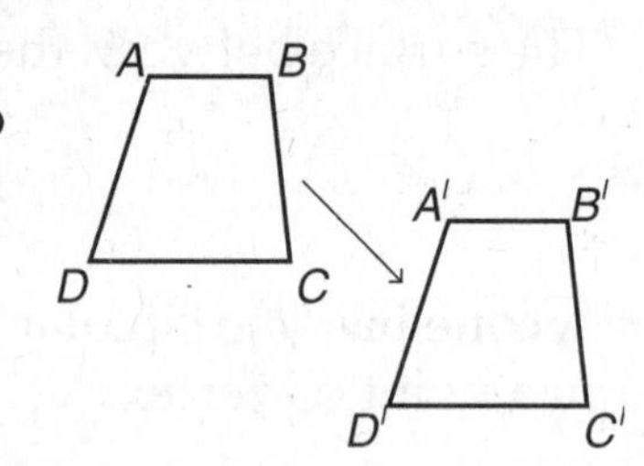

A rotation or turn is a turn a certain number of degrees around a fixed point. If you rotate a figure 360°, it comes back to its starting position. To find whether a rotation has been performed, you can trace the figure and the fixed point. Keep the fixed points lined up and turn the paper to see whether the traced figure lines up with the second image.

More than one transformation can be performed on a shape. For example, trapezoid $W'X'Y'Z'$ is the image of $WXYZ$ after being slid over the line and then rotated clockwise 90° around point P.

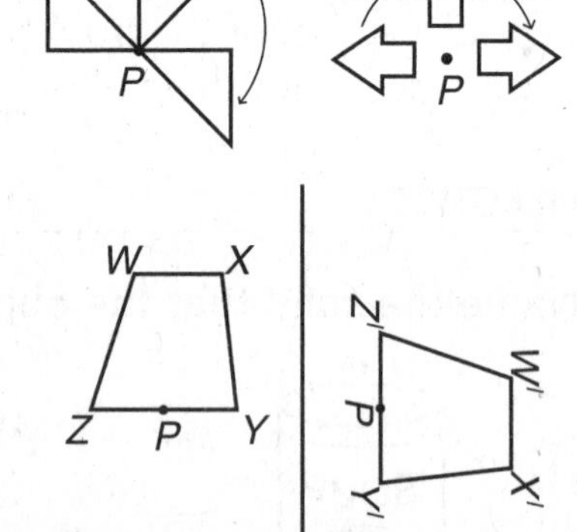

PRACTICE

Circle the letter of the correct name for the transformation that is shown.

1.

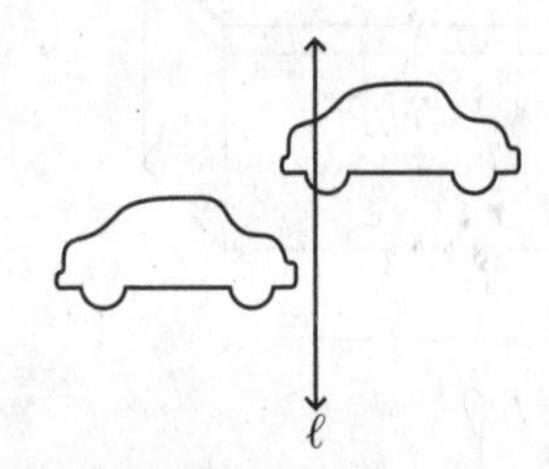

A reflection
B rotation
C translation

2.

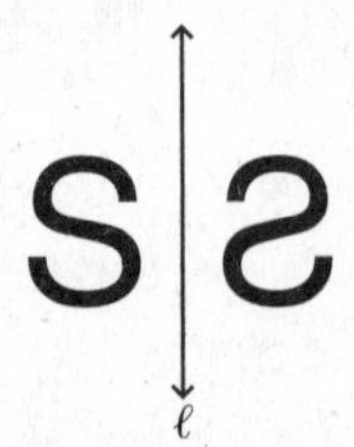

F reflection
G rotation
H translation

3.

A reflection
B rotation
C translation

4.

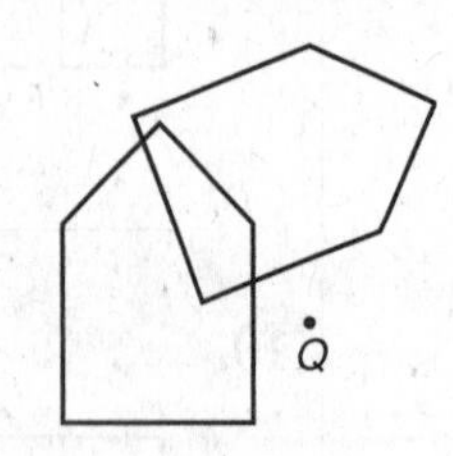

F reflection
G rotation
H translation

5.

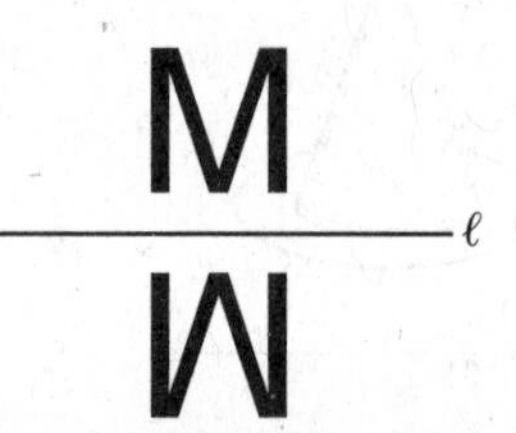

A reflection
B rotation
C translation

6.

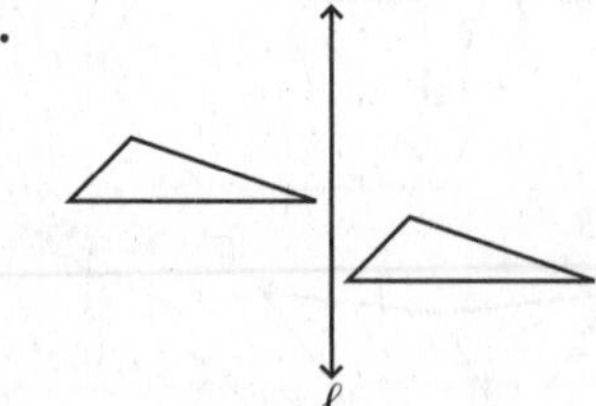

F reflection
G rotation
H translation

For Numbers 7–9, tell how many degrees the figure has been rotated.

7.

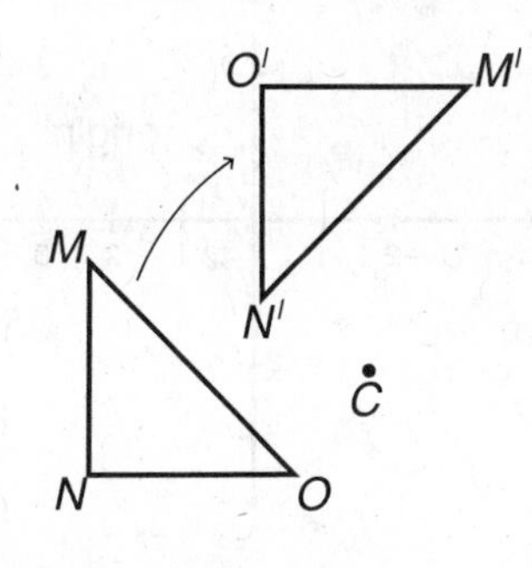

____

8.

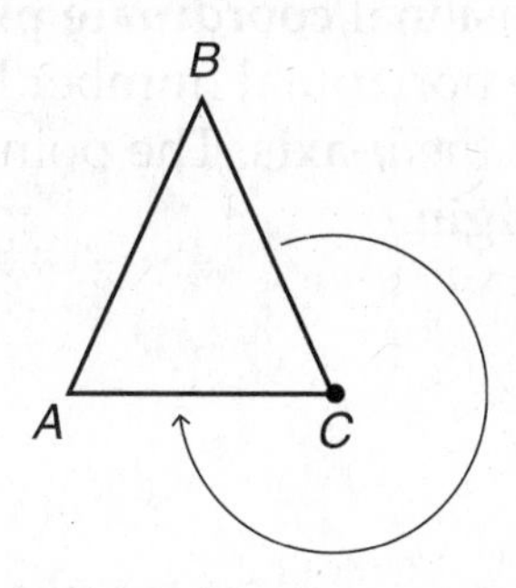

____

9.

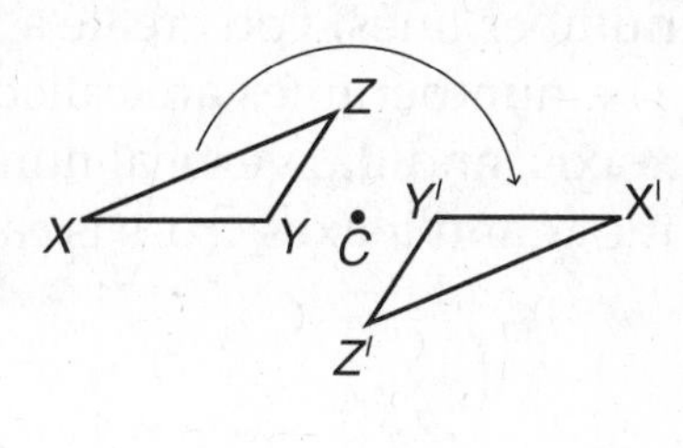

____

For Numbers 10–11, circle the letter of the steps taken to transform Figure 1 to Figure 2.

10.

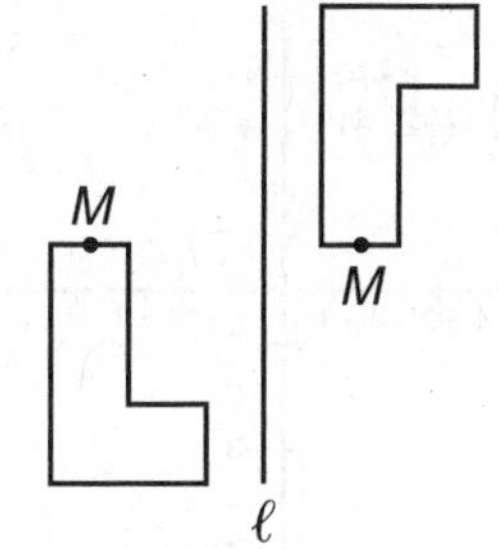

Figure 1 Figure 2

A Reflect Figure 1 across line *l*, and rotate it clockwise 90° about Point *M*.

B Reflect Figure 1 across line *l*, but do not rotate it.

C Reflect Figure 1 across line *l*, and rotate it clockwise 180° about Point *M*.

D Reflect Figure 1 across line *l*, and rotate it counterclockwise 90° about Point *M*.

11.

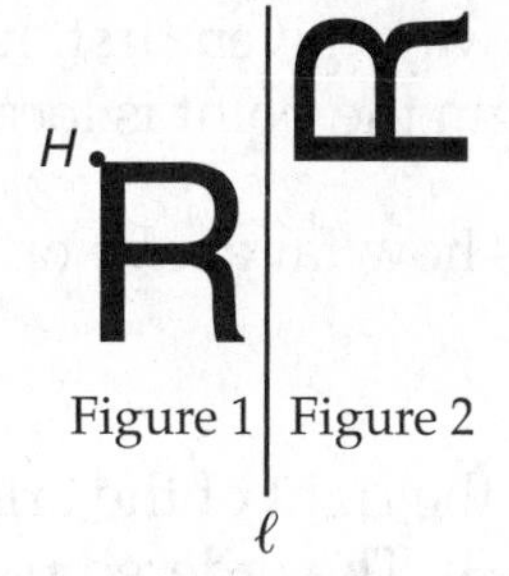

Figure 1 Figure 2

ℓ

F Slide Figure 1 across line *l*, and rotate it clockwise 90° about Point *H*.

G Slide Figure 1 across line *l*, and rotate it counterclockwise 180° about Point *H*.

H Slide Figure 1 across line *l*, but do not rotate it.

J Slide Figure 1 across line *l*, and rotate it clockwise 270° about Point *H*.

For Number 12, circle the letter of the figure that illustrates quadrilateral *WXYZ* reflected over $\overline{WZ}$.

12.

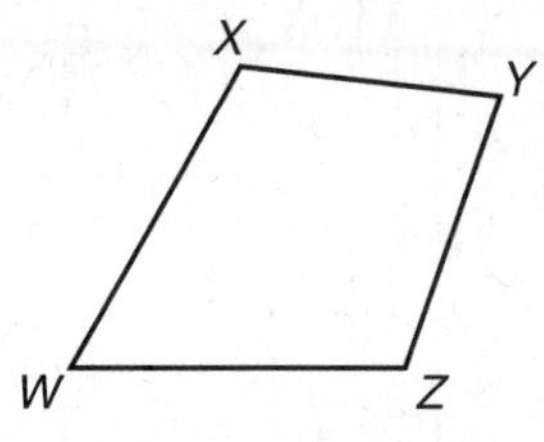

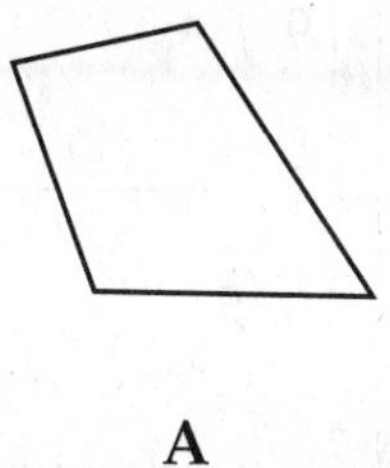

A

B

C

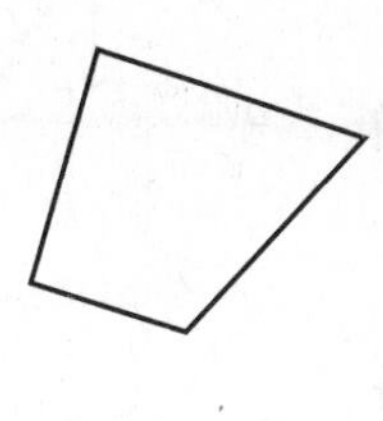

D

Locating Points on the Coordinate Plane

When you divide a **plane**, or flat surface, with a pair of perpendicular number lines, you create a two-dimensional **coordinate plane**. The number lines are called **axes**. The horizontal number line is the ***x*-axis**, and the vertical number line is the ***y*-axis**. The point where the *x*- and *y*-axes cross is called the **origin**.

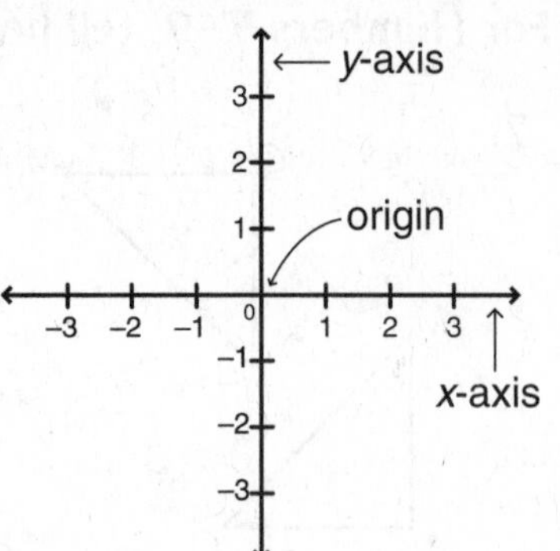

The coordinate plane provides a useful system for locating or giving the location of a point. Maps are created using such a system. A location is given by two numbers called **coordinates** written in parentheses and separated by a comma. Together, the coordinates form an **ordered pair**.

- The ***x*-coordinate** is always given first. It tells how far to the left or right of the origin the point is located.
- The ***y*-coordinate** tells how far above or below the origin the point is located.

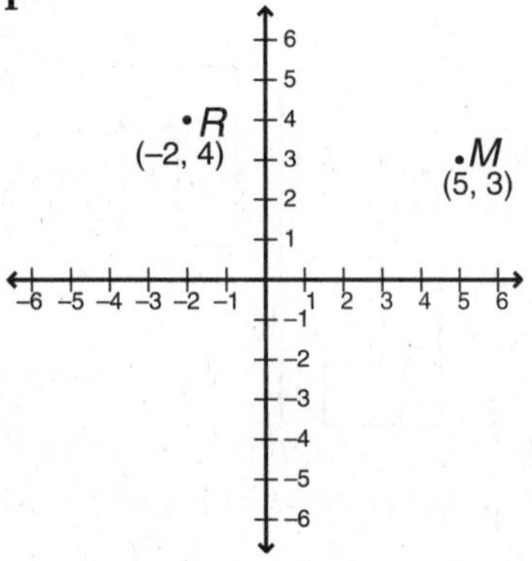

At right, *M* is 5 units to the right of the origin and 3 units up, so its ordered pair is (5, 3). The ordered pair for point *R* is (-2, 4) because *R* is 2 units to the left of the origin and 4 units up.

PRACTICE

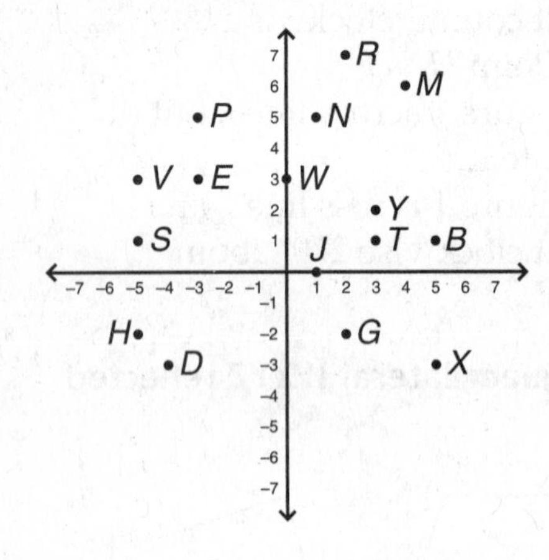

Write the letter for each coordinate pair.

1. (3, 2) ________
2. (5, 1) ________
3. (1, 5) ________
4. (−3, 3) ________
5. (−5, 1) ________
6. (−5, −2) ________

Write the ordered pair for each point.

7. *G* ________
8. *M* ________
9. *J* ________
10. *R* ________
11. *X* ________
12. *W* ________

Geometry Skills Checkup

Circle the letter for the correct answer to each question.

Study the diagram below. Then do Numbers 1–6.

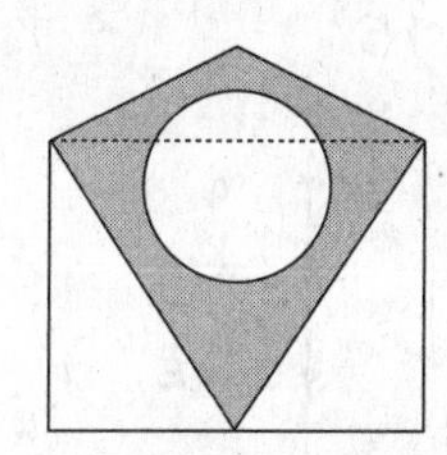

1. What shape is formed by the outline of this diagram?
   - **A** a pentagon
   - **B** an octagon
   - **C** a hexagon
   - **D** a quadrilateral

2. Which parts of this diagram, if any, are congruent?
   - **F** the shaded shape and the outside edges
   - **G** the circle and the shaded shape
   - **H** the two unshaded triangles
   - **J** No two of the shapes are congruent.

3. What types of triangles are shown in the bottom corners of the diagram?
   - **A** equilateral
   - **B** isosceles
   - **C** right
   - **D** obtuse

4. What type of angle is shown at the very top of the diagram?
   - **F** an acute angle
   - **G** a right angle
   - **H** a square corner
   - **J** an obtuse angle

5. Which line segments in the diagram, if any, are parallel?
   - **A** the bottom and the right side
   - **B** the top and the left side
   - **C** the right and left sides
   - **D** There are no parallel line segments in the diagram.

6. Which type of polygon is formed by the outline of the shaded area?
   - **F** a regular polygon
   - **G** a quadrilateral
   - **H** a parallelogram
   - **J** all of the above

7. How many different diagonals can be drawn to divide this figure in half?

   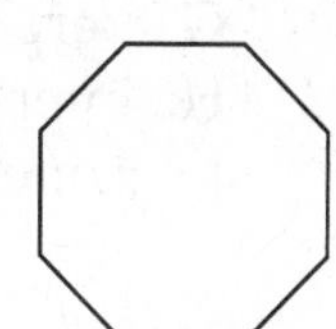

   - **A** 2
   - **B** 4
   - **C** 6
   - **D** 8

8. Which pattern could be folded to make the figure shown at the right?

   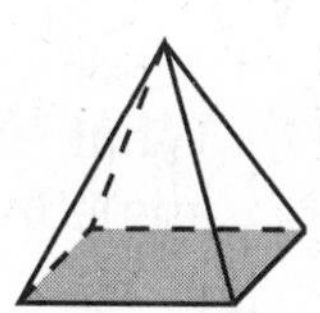

   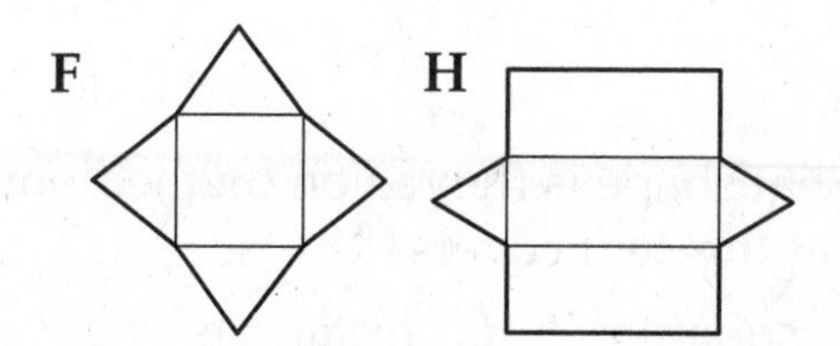

   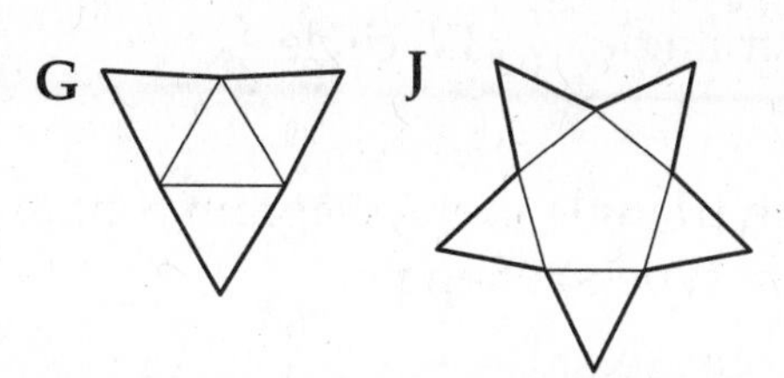

Geometry Skills Checkup (continued)

The diagram below shows 4 fabric pieces for the tent shown at the right. The sides of the tent are congruent triangles. Study the diagram. Then do Numbers 9–13.

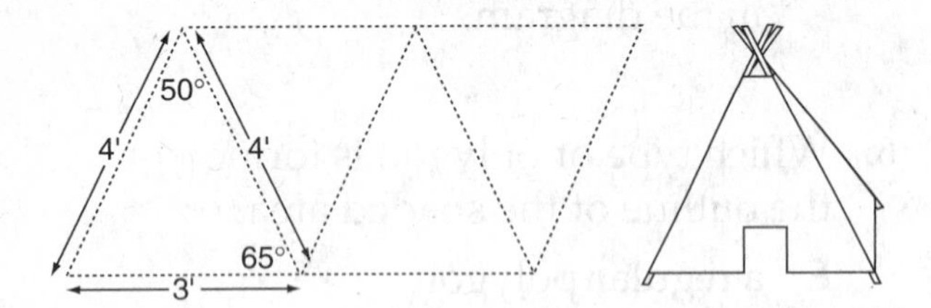

9. What type of line segments are formed by the tent poles?
   - F parallel
   - G perpendicular
   - H intersecting
   - J none of the above

10. What kind of triangle is formed by each side of the tent?
    - A equilateral
    - B isosceles
    - C right
    - D obtuse

11. What is the measure of the unmarked angle in the first triangle?
    - F 50°
    - G 100°
    - H 60°
    - J 65°

12. What shape is the region on the floor that this tent covers?
    - A square
    - B triangle
    - C rectangle
    - D circle

13. Each triangle in the diagram contains what type(s) of angles?
    - F obtuse only
    - G obtuse and acute
    - H acute and right
    - J acute only

Use the coordinate grid below to answer Numbers 14 and 15.

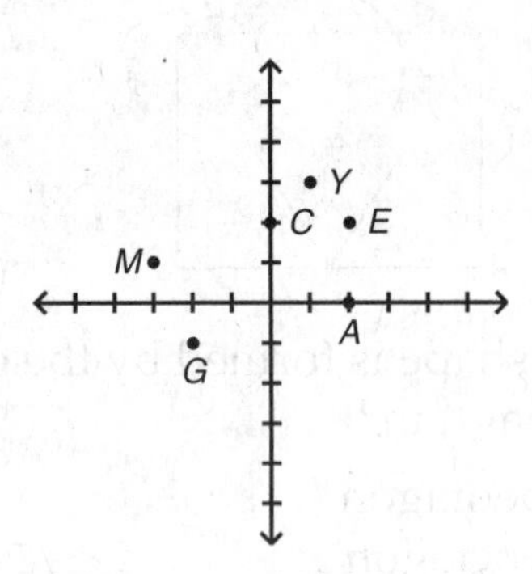

14. Which ordered pair names the location of Point *A*?
    - A (2, 2)
    - B (-2, 2)
    - C (0, 2)
    - D (2, 0)

15. Which point is located at (1, 3)?
    - F Point *C*
    - G Point *M*
    - H Point *Y*
    - J Point *G*

16. This line segment names a diameter of the circle.

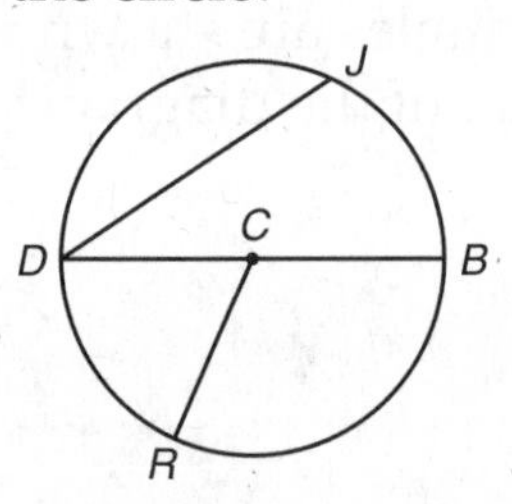

- A $\overline{CR}$
- B $\overline{BJ}$
- C $\overline{DB}$
- D $\overline{DC}$

17. Which type of transformation is shown in the diagram?
    - F reflection
    - G rotation
    - H translation
    - J none of the above

K | K

Measurement

Reading a Scale

A number line on a measurement tool is called a **scale**. To use a simple scale such as a ruler, follow these steps:

- Line up one end of the object you are measuring with the zero.
- Find the number that lines up with the other end of the object.
- Check your work by measuring the object again.

Sometimes a scale skips numbers. In that situation, you can estimate the measurement shown on the scale. The pointer is about halfway between 10 and 20, so the measurement shown on the scale is about 15.

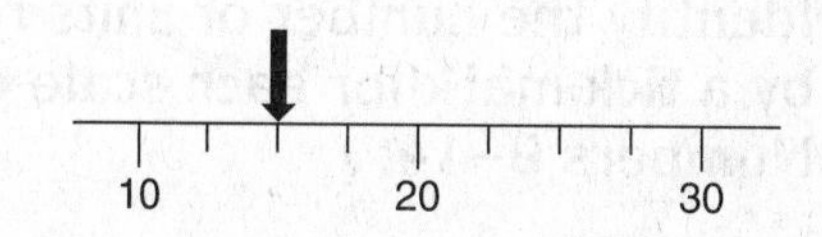

PRACTICE

Find each measurement below.

1. The pointer on a scale is about halfway between 18 and 20. What number is halfway between 18 and 20? ________

2. The pointer on a scale is about halfway between 90 and 100. What number is halfway between 90 and 100? ________

3. The pointer on a scale is about halfway between 60 and 80. What number is halfway between 60 and 80? ________

4. The pointer on a scale is about $\frac{1}{4}$ of the way from 60 to 80. What measurement is shown? ________

5. The pointer on a scale is about $\frac{1}{4}$ of the way from 0 to 100. What measurement is shown? ________

6. The pointer on a scale is about $\frac{1}{3}$ of the way from 60 to 90. What measurement is shown? ________

7. What measurement is shown on this scale? ________

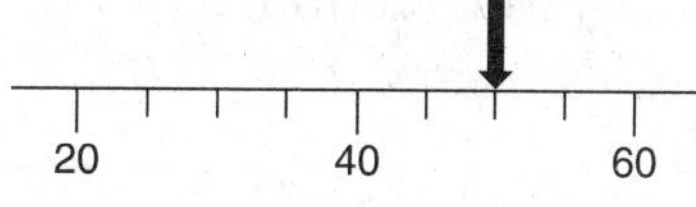

8. What measurement is shown on this scale? ________

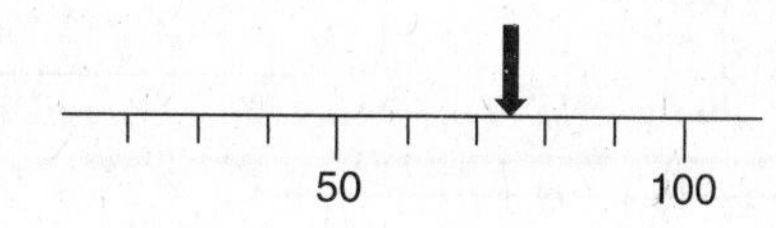

Reading a Scale (continued)

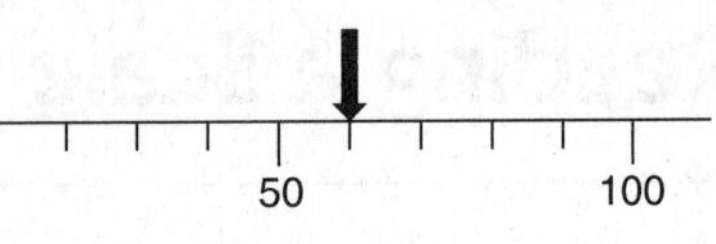

If a scale has tick marks, count the number of equal spaces formed between numbers. Divide the difference between the numbers by the number of spaces to find the value of a tick mark. In the scale shown, there are 5 equal spaces between 50 and 100. $100 - 50 = 50$, and $50 \div 5 = 10$. Each tick mark represents 10 units.

The arrow is one tick mark beyond 50, so the reading is $50 + 10$, or 60 units.

PRACTICE

Identify the number of units represented by a tick mark for each scale described in Numbers 9–14.

9. The interval between 30 and 40 is divided into 5 equal parts. Each tick mark represents ________.

10. The interval between 60 and 80 is divided into 10 equal parts. Each tick mark represents ________.

11. The interval between 75 and 100 is divided into 5 equal parts. Each tick mark represents ________.

12. The interval between 100 and 200 is divided into 5 equal parts. Each tick mark represents ________.

13.
0 100

Each tick mark represents ________.

14.
20 40

Each tick mark represents ________.

For Numbers 15–18, identify the value of each tick mark and the reading on the scale.

15.
0 100

Each tick mark represents ________.
The reading on the scale is ________.

16.
30 40

Each tick mark represents ________.
The reading on the scale is ________.

17.
0 50

Each tick mark represents ________.
The reading on the scale is ________.

18.
40 50

Each tick mark represents ________.
The reading on the scale is ________.

Using a Customary Ruler

In the **customary system**, most standard rulers, yardsticks, and tape measures are not as simple as the scales on the previous pages. The inch ruler shown here uses several different types of tick marks.

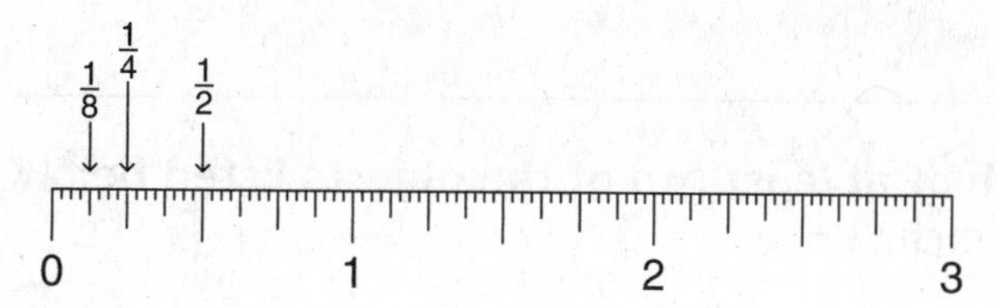

- The longest tick marks divide the ruler into inches.
- The next longest tick marks divide each inch into half inches.
- The half inches are divided into fourths by slightly shorter tick marks.
- The fourths are divided into eighths by even shorter tick marks.
- The shortest tick marks may be eighths of an inch or sixteenths of an inch.

PRACTICE

Use a ruler marked in inches. Measure each line below to the nearest $\frac{1}{2}$ inch. Some answers will be whole numbers.

1. ______________________

   _______ inches

2. ______________________________

   _______ inches

3. _______

   _______ inch

4. ____________________________________

   _______ inches

5. _____________

   _______ inch

6. ________________________

   _______ inches

7. _________

   _______ inch

Measure the lines below to the nearest $\frac{1}{4}$ inch.

8. ___

   _______ inch

9. _________

   _______ inch

10. __________________

    _______ inches

11. _______

    _______ inch

12. _________________________

    _______ inches

13. _____________

    _______ inches

14. ______________________________

    _______ inches

Measuring to a Fraction of an Inch

When you measure with a ruler, make sure you line up one end of the object with 0. Then find the tick mark that lines up with the other end of the object. To complete the items on this page, you will measure objects in your home or your classroom using a customary ruler.

PRACTICE

Find the height and width of at least *two* of the objects listed below. Give your answers to the nearest fourth of an inch.

1. a VCR tape _____ by _____
2. a new #2 pencil _____ by _____
3. a standard index card _____ by _____
4. the front of this book _____ by _____

The objects listed below are difficult to measure because they have rounded shapes. Measure the widest and tallest part of each object. Give your answers to the nearest half of an inch.

5. a 12-oz soda can _____ by _____
6. a 2-liter soda bottle _____ by _____

Measure the height and width of at least *two* of the objects listed below. Give your answers to the nearest eighth of an inch.

7. a can of Campbell's® condensed soup _____ by _____
8. a computer disk _____ by _____
9. a dollar bill _____ by _____
10. a 9-volt battery _____ by _____

Circle the letter for the most reasonable measurement of each object.

11. the width of a dime

A $\frac{11}{16}$ inch C $\frac{1}{2}$ inch

B $\frac{3}{4}$ inch D $\frac{7}{8}$ inch

12. the width of a penny

F $\frac{4}{8}$ inch H $\frac{3}{4}$ inch

G $\frac{7}{8}$ inch J $\frac{14}{16}$ inch

13. the width of a nickel

A 1 inch C $\frac{1}{4}$ inch

B $\frac{5}{8}$ inch D $\frac{13}{16}$ inch

Circle the letter for the most accurate measurement of each line below.

14. —

F $\frac{1}{2}$ inch H $\frac{1}{16}$ inch

G $\frac{5}{16}$ inch J $\frac{3}{16}$ inch

15. ———

A $\frac{3}{4}$ inch C $\frac{3}{8}$ inch

B $\frac{7}{16}$ inch D $\frac{5}{8}$ inch

16. ——————————

F $\frac{3}{4}$ inch H $1\frac{3}{8}$ inches

G 1 inch J $\frac{1}{2}$ inch

Reviewing Customary Units of Measure

The customary system of measurement is the one most commonly used in the United States.

Customary System Units of Measure

Temperature degrees Fahrenheit (°F)	Normal body temperature is 98.6°F. Water boils at 212°F. Water freezes at 32°F.
Length 12 inches (in.) = 1 foot (ft) 3 feet = 1 yard (yd) 5,280 feet = 1 mile (mi)	An inch is about the length of a straight pin. A foot is about the length of a man's foot. A yard is about the length of your arm. A mile is 8 to 10 city blocks.
Weight 1 pound (lb) = 16 ounces (oz) 1 ton (T) = 2,000 lb	A pencil weighs about 1 ounce. An eggplant weighs about 1 pound. A car weighs about 1 ton.
Capacity 1 pint = 2 cups 1 quart = 4 cups (or 2 pints) 1 gallon = 4 quarts (or 16 cups)	An ice cream dish holds about 1 cup. A mug holds about 1 pint. A narrow milk carton holds 1 quart. A large plastic milk jug holds 1 gallon.

Time
1 minute (min) = 60 seconds (sec)
1 hour (hr) = 60 min
1 day = 24 hr

PRACTICE

Circle the letter for the most reasonable estimate for each measurement.

1. the temperature in a refrigerator
   - A 70°F
   - B 25°F
   - C 40°F

2. the width of a dime
   - F 3 inches
   - G $\frac{1}{2}$ inch
   - H 6 inches

3. the weight of a dinner roll
   - A 1 pound
   - B 1 ounce
   - C $\frac{1}{2}$ pound

4. the capacity of a soup can
   - F $1\frac{1}{2}$ cups
   - G 1 gallon
   - H 1 quart

Circle the *greater* measurement in each pair.

5. 15 inches 1 foot
6. 5 feet 2 yards
7. 12 ounces 1 pound
8. 22 ounces 1 pound
9. 3 cups 1 pint
10. 3 cups 1 quart
11. 1 gallon 12 cups
12. 90 minutes 2 hours

Converting Units Within the Customary System

In order to convert from one unit of measurement to another, you need to know the exchange rate. For example, in changing from feet to inches, you need to know that 1 foot = 12 inches. You can use the table on the previous page as a reference.

One way to convert between units is to set up a proportion. Be sure to place like units in the numerator and like units in the denominator.

Example Change 15 feet to yards.
To make this conversion, you need to know that 3 feet = 1 yard.

- Set up a proportion. $\frac{\text{feet}}{\text{yards}}$ $\frac{3}{1} = \frac{15}{n}$
- Cross multiply. $3n = 15$
- Divide both sides by 3. $\frac{\cancel{3}^{1}n}{\cancel{3}_{1}} = \frac{\cancel{15}^{5}}{\cancel{3}_{1}}$

15 feet = 5 yards $n = 5$

Another way to convert between units is to use multiplication or division.
- Multiply to change larger units to smaller units.

Example 1 foot = 12 inches To convert from feet to inches, multiply each foot by 12.

- Divide to change smaller units to larger units.

Example 3 feet = 1 yard To convert from feet to yards, divide the number of feet by 3.

PRACTICE

Convert each measurement below to the given unit.

1. 9 ft = ________ yd
2. 24 in. = ________ ft
3. 36 in. = ________ yd
4. $\frac{1}{2}$ ft = ________ in.
5. $\frac{1}{3}$ yd = ________ ft
6. $1\frac{1}{2}$ feet = ________ in.
7. $1\frac{2}{3}$ yd = ________ ft
8. 3 lb = ________ oz
9. $\frac{1}{2}$ lb = ________ oz
10. 32 oz = ________ lb
11. 2 tons = ________ lb
12. 360 min = ________ hr
13. $1\frac{1}{2}$ hr = ________ min
14. $\frac{1}{4}$ hour = ________ min
15. 1 ft 2 in. = ________ in. (Convert 1 foot into inches and then add 2 in.)
16. 11 ft = _____ yd _____ ft (Divide 11 by 3 and give the remainder in feet.)
17. 23 oz = _____ lb _____ oz (Divide 23 by 16 and give the remainder in ounces.)

Reviewing Metric Units of Measure

Most countries around the world use the metric system of measurement.

Metric Units of Measure

Temperature degrees Celsius (°C)	Water boils at 100°C. Water freezes at 0°C. Normal body temperature is 37°C.
Length – basic unit is the meter 1 meter (m) = 1,000 millimeters (mm) = 100 centimeters (cm) 1 cm = 10 mm 1 kilometer (km) = 1,000 m	A needle is about 1 millimeter wide. A kindergarten student is about 1 meter tall. Your little finger is about 1 centimeter wide.
Mass (Weight) – basic unit is the gram 1 gram (g) = 1,000 milligrams (mg) 1 kilogram (kg) = 1,000 g	A needle weighs about 1 milligram. A peanut weighs about 1 gram. A city telephone book weighs about 1 kg.
Capacity – basic unit is the liter 1 liter (L) = 1,000 milliliters (mL) 1 kiloliter (kL) = 1,000 L	A large plastic soda bottle holds 2 liters. A dose of cough medicine is about 10 mL. A septic tank holds about 2 kiloliters.

The rate of exchange for the metric system corresponds to that of our decimal place-value system. When you get 10 units, you regroup. Prefixes identify amounts less than or greater than the basic units. The most frequently used prefixes and their abbreviations are shown below.

1,000	100	10	1	0.1 *or* $\frac{1}{10}$	0.01 *or* $\frac{1}{100}$	0.001 *or* $\frac{1}{1,000}$
kilo (k)	hecto (h)	deka (dk)	basic unit	deci (d)	centi (c)	milli (m)
km kL kg	hm hL hg	dkm dkL dkg	meter (m) liter (L) gram (g)	dm dL dg	cm cL cg	mm mL mg

PRACTICE

Circle the letter of the most reasonable measure.

1. The capacity of a coffee mug — **A** 2.5 kL — **B** 25 L — **C** 250 mL
2. The oil needed to fry an egg — **D** 5 L — **E** 5 dL — **F** 5 mL
3. The length of a new pencil — **G** 14 cm — **H** 14 mm — **J** 14 km
4. The mass of a cat — **A** 20 g — **B** 200 dg — **C** 2 kg
5. The length of a baseball bat — **D** 1 dkm — **E** 1 m — **F** 1 dm
6. The mass of a wristwatch — **G** 30 mg — **H** 30 g — **J** 30 kg

Converting Units Within the Metric System

As with customary units, you can use a proportion to convert between units within the metric system. You must use the rate of exchange between the units to set up the proportion. In the metric system, the converted units should be expressed as decimals rather than fractions.

Example 24 cg = _____ g

- Set up a proportion. $\frac{\text{cg}}{\text{g}} \quad \frac{100}{1} = \frac{24}{n}$
- Cross multiply. $100 \times n = 24 \times 1$
- Divide both sides by 100. $\frac{100n}{100} = \frac{24}{100} = 0.24$

24 cg = 0.24 g

Another way to convert between units is to use multiplication or division. You can use a place-value chart to help. (See the place-value chart on page 135.)

- Multiply to convert larger units to smaller units.

Example 5 hm = _____ dm

Find the **hecto**meters in the chart. **Deci**meters are three columns to the right. Each column has ten times the value of the column immediately to its right. When you move to the right, multiply. For three columns to the right, multiply by $10 \times 10 \times 10$, or 1,000. To convert hectometers to decimeters, multiply the number of hectometers by 1,000.

$5 \times 1{,}000 = 5{,}000$ 5 hm = 5,000 dm

- Divide to convert smaller units to larger units.

Example 14 dL = _____ kL

Find the **deci**liters column in the chart. **Kilo**liters are four columns to the left. For each column to the left, divide. For four columns to the left, divide by $10 \times 10 \times 10 \times 10$, or 10,000. Remember that in the metric system, units are always expressed as decimals.

$14 \div 10{,}000 = 0.0014$ 14 dL = 0.0014 kL

PRACTICE

Fill in each blank.

1. $\frac{1}{2}$ kg = _____ g
2. 2 L = _____ mL
3. 3,000 mL = _____ L
4. $1\frac{1}{2}$ m = _____ mm
5. 3,500 mm = _____ m
6. 10 cm = _____ mm
7. $2\frac{1}{2}$ kL = _____ L
8. 150 cm = _____ m
9. 3 m 15 cm = _____ cm
10. 1 kg 450 g = _____ g
11. 3 L 15 mL = _____ mL
12. 1,515 g = _____ kg _____ g

Comparing Customary and Metric Units

The chart below compares some customary units with metric units of measurement. The symbol $\approx$ indicates that the measurements are approximately equal.

Length	Mass/Weight	Capacity
1 in. $\approx$ 2.5 cm 1 m $\approx$ 1.1 yd 1 km $\approx$ 0.6 mi	1 kg $\approx$ 2.2 lb 1 oz $\approx$ 28 g	1 L $\approx$ 1.1 qt 1 kL $\approx$ 275 gal

You can use a proportion to convert a measurement from one system to the other. Use the exchange rates shown in the chart above to set up the proportion.

Example 40 cm $\approx$ ______ in.

- Set up the ratio. $\frac{\text{cm}}{\text{in.}}$ $\frac{2.5}{1} \approx \frac{40}{n}$
- Cross multiply. $2.5n \approx 40$
- Divide both sides by 2.5 $\frac{2.5n}{2.5} \approx \frac{40}{2.5}$

$n \approx 16$

40 cm $\approx$ 16 in.

PRACTICE

Circle the greater measurement in each pair.

1. 1 in. 1 cm
2. 1 g 1 oz
3. 1 qt 1 L
4. 1 yd 1 m
5. 1 mi 1 km

Convert units as indicated.
Round all answers to the nearest tenth.

6. 20 yd $\approx$ ______ m
7. 40 g $\approx$ ______ oz
8. 15 mi $\approx$ ______ km
9. 10 L $\approx$ ______ qt
10. 3 m $\approx$ ______ yd
11. 15 lb $\approx$ ______ kg
12. 100 cm $\approx$ ______ in.
13. 150 gal $\approx$ ______ kL
14. 12 in. $\approx$ ______ cm
15. 4 km $\approx$ ______ mi

Adding and Subtracting Mixed Measurements

To add or subtract mixed measurements, start computing with the smallest unit, and then add or subtract the next greatest unit. You may need to regroup. When that happens, be sure you use the exchange rate that corresponds to the units of measure you are working with. Simplify answers.

Examples

Add 1 hour 15 minutes and 2 hours 50 minutes.

- Add minutes first.
- Then add hours.

1 hour	15 minutes
+ 2 hours	50 minutes
3 hours	65 minutes

- Simplify. Regroup minutes to hours. Remmeber 60 minutes = 1 hour.

3 hours + 65 minutes = 3 hours + 60 minutes + 5 minutes
= 3 hours + 1 hour + 5 minutes
= 4 hours 5 minutes

Subtract 1 foot 9 inches from 3 feet 3 inches.

- Subtract inches first.
- Continue subtracting.

~~3~~ (2) feet	~~3~~ (15) inches
− 1 foot	9 inches
1 foot	6 inches

Since 9 inches is greater than 3 inches, change 1 foot to 12 inches and add to the 3 inches.

PRACTICE

Add or subtract. Then simplify your answer. You can use the table on page 133.

1. 2 hours 45 minutes
+ 1 hour 15 minutes

2. 4 pounds 10 ounces
+ 5 pounds 7 ounces

3. 5 yards 2 feet
+ 2 yards 2 feet

4. 12 feet 10 inches
+ 3 feet 7 inches

5. 1 hour 15 minutes
+ 50 minutes

6. 1 quart 3 cups
+ 1 quart 2 cups

7. 12 pounds 10 ounces
− 5 pounds 7 ounces

8. 3 yards 2 feet
− 2 yards 1 foot

9. 9 feet 4 inches
− 5 feet 8 inches

10. 1 gallon 1 quart
− 3 quarts

11. 4 hours 15 minutes
− 1 hour 45 minutes

12. 4 pints
− 1 pint 1 cup

Calculating Time

Sometimes you need to figure out when to start a task in order to complete it by a certain time. Or you might need to know what time you will finish a task if you start it at a certain time. To calculate this, first count the complete number of hours, then count the number of minutes.

Examples

It takes 1 hour 15 minutes to drive to the doctor's office. You have a 10:00 appointment. If you allow an additional 15 minutes for traffic problems, what time should you leave?

Hint: You need time before 10:00, so count back from 10:00.

- Count the hours. 1 hour before 10:00 is 9:00.
- Count the minutes. 15 minutes plus 15 additional minutes equals 30 minutes. 30 minutes before 9:00 is 8:30.

You should leave at 8:30.

Wendi put a cake into the oven at 8:15. The cake is supposed to bake for 40 minutes. What time should the cake be ready to come out of the oven?

- Count the hours. There are only minutes to count.
- Count the minutes. 15 minutes plus 40 minutes equals 55 minutes. Add the minutes to the hours. 8 hours + 55 minutes = 8:55

The cake should be ready at 8:55.

Pay close attention to any mention of A.M. (morning) and P.M. (afternoon and evening). From 9:00 A.M. to 10:00 A.M. is 1 hour, but from 9:00 A.M. to 10:00 P.M. is 13 hours.

PRACTICE

Fill in the blank for each problem.

1. What time is it 3 hours *before* 1:15?

______

2. It will take $3\frac{1}{2}$ hours to make dinner rolls. You plan to eat at 12:15 P.M. What time should you start making the rolls?

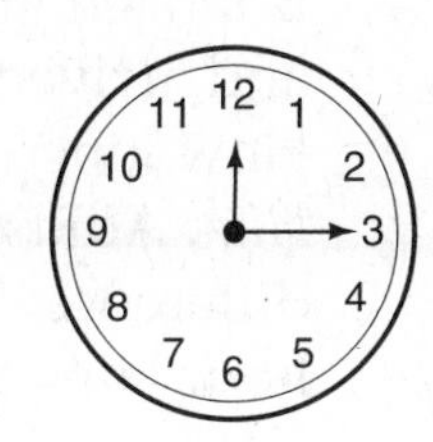

______

3. You need to work $5\frac{1}{2}$ hours today. If you start at 11:15, what time will you finish? ______

4. Your night class starts at 6:55 and ends at 7:35. How long is the class? ______

5. It takes you 20 minutes to clean 1 room, and there are 3 rooms in your apartment. If you start cleaning at 8:10, what time will you finish? ______

Finding Elapsed Time

Elapsed time is the amount of time that passes from the start of an event or activity to the end of that event or activity.

Example

Will started painting the kitchen at 2:30 in the afternoon. He finished at 9:15 that evening. How long did it take Will to paint the kitchen?

One method to solve is to subtract.
- Begin with the finish time and subtract the start time.
- Regroup as needed.

```
   8    75
   9 hr 15 min   (9 and 15 crossed out)
 − 2 hr 30 min
 -------------
   6 hr 45 min
```

Another method is to count forward.
- Begin with the start time and count the hours to the finish time.

  From 2 to 9 is 7 hours.
- Adjust for the minutes.

  Subtract 30 minutes for the time from 2 to 2:30.

  Add on 15 minutes for the time from 9 to 9:15.

```
   6 hr 60 min
 −      30 min
 -------------
   6 hr 30 min
 +      15 min
 -------------
   6 hr 45 min
```

PRACTICE

Find the elapsed time for each situation.

1. When Clyde's train left the station, it was 5:17 A.M. When the train arrived at his destination, it was 6:39 A.M. How long was Clyde's train ride? ________

2. Jennifer got back from lunch at 1:45. When she left for lunch, it was 11:20. How much time did Jennifer take for lunch? ________

3. The concert started at 8:30 P.M. and ended at 10:10 P.M. How long did the concert last? ________

4. Paul started working on his model plane at 10:30 A.M. At 3:15 P.M. he stopped working to have lunch. How long had Paul been working on his model? ________

5. Celine ran in the marathon. The race began at 9:00 A.M. Celine crossed the finish line at 1:15. How long did Celine's run last? ________

6. The inspection crew started at 11:30 A.M. and completed their work at 2:20 P.M. How long did the job take? ________

7. Brian's flight left Chicago at 10:00 A.M. and arrived in New York at 1:09 P.M. How long was the flight? *Hint:* Adjust for the time difference. New York is one hour later than Chicago. ________

8. The first guest arrived for Connie's party at 8:00 P.M. The last guest left at 2:30 A.M. How long did the party last? ________

Finding Perimeter

Perimeter is a measure of the distance around the outside edge of a figure. To find the perimeter, add the lengths of the figure's sides together. A capital letter P is used to represent perimeter.

Opposite sides of a rectangle are the same length. To find the perimeter of a rectangle, add the lengths of the four sides.

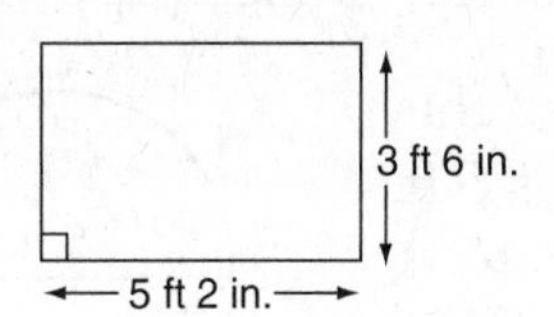

$$\begin{array}{r} 5 \text{ ft } 2 \text{ in.} \\ 3 \text{ ft } 6 \text{ in.} \\ 5 \text{ ft } 2 \text{ in.} \\ + \; 3 \text{ ft } 6 \text{ in.} \\ \hline 16 \text{ ft } 16 \text{ in.} \end{array}$$

$P = 17$ ft 4 in.

The dashed lines in the figure at the right show that the figure fits inside a rectangle. The top and bottom of the rectangle are each 2 m. The right side is $1\frac{1}{2}$ m, and the sum of the two sides along the left is also $1\frac{1}{2}$ m.

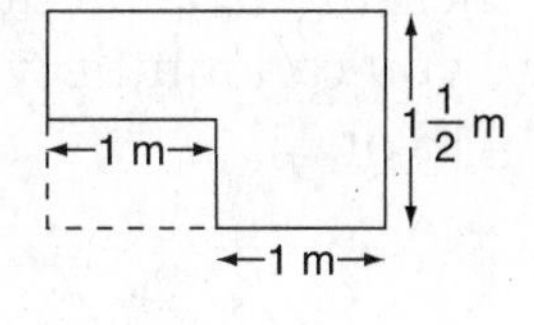

$$\begin{array}{r} 2 \text{ m} \\ 1\frac{1}{2} \text{ m} \\ 2 \text{ m} \\ + \; 1\frac{1}{2} \text{ m} \\ \hline \end{array}$$

$P = 6\frac{2}{2}$ m $=$ 7m

PRACTICE

Find the perimeter of each shape.

1. 10 ft, 26 ft, 24 ft

   $P =$ ____________

2. 3 cm, 3 cm, 3 cm, 3 cm, 3 cm, 3 cm

   $P =$ ____________

3. 10 in., 5 in., 3 in., 6 in.

   $P =$ ____________

4. 4 ft 2 in., 7 ft 3 in.

   $P =$ ____________

5. 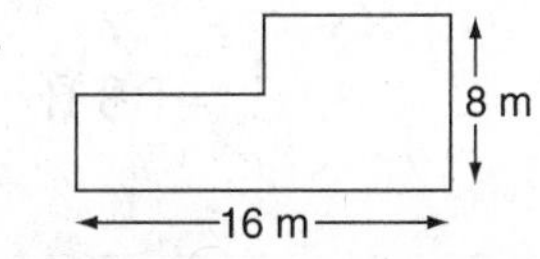

   $P =$ ____________

6. 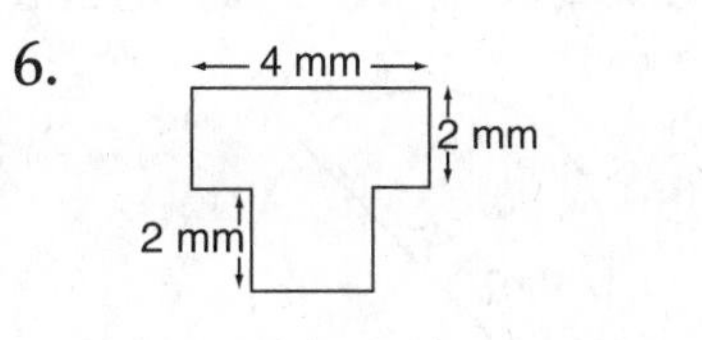

   $P =$ ____________

7. 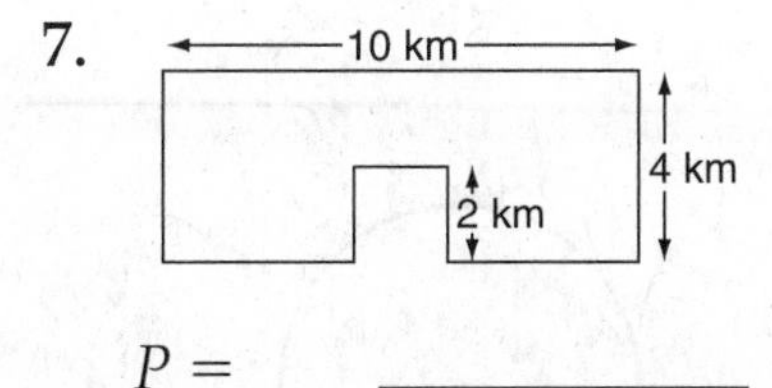

   $P =$ ____________

8. 8 in., w

   $P = 24$ in.

   $w =$ ____________

9. What is the perimeter of a room 15 ft long and 12.5 ft wide?

   $P =$ ____________

10. How much fencing is needed to enclose a 6 × 8 foot garden?

    $P =$ ____________

11. If the perimeter of a square is 16 inches, how long is each side of the square?

    $P =$ ____________

12. If the perimeter of an equilateral triangle is 54 cm, how long is each side of the triangle?

    $P =$ ____________

Finding Circumference

The perimeter of a circle is called the **circumference.**
A capital letter C is used to represent circumference.

The circumference of any circle is approximately 3 times its diameter (d).

- The diameter (d) of the circle at the right is 5 ft. $C \approx 3 \times 5$ ft, or ≈ 15 ft

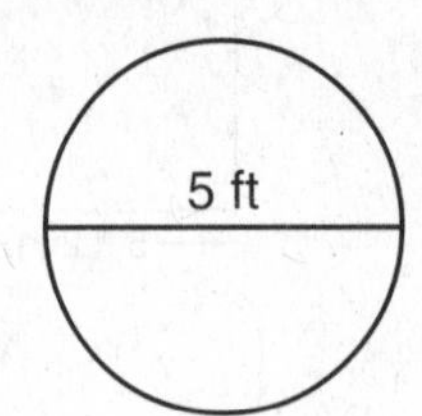

- The radius (r) of the circle at the right is 6 yd. The diameter is double the radius, or 12 yd. $C \approx 3 \times 12$ yd, or ≈ 36 yd

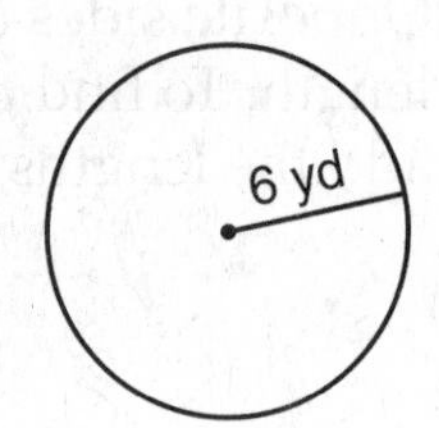

To find a more exact measure of a circumference, instead of saying "it is about 3 times the diameter," you can multiply the diameter by the Greek symbol π, which represents 3.1416 or $\frac{22}{7}$.

Examples

Find the circumference of a circle with a radius of 4 cm. Use $\pi = 3.14$.

$C = \pi \times d$

$r = 4$ cm, so $d = 8$ cm

$C = 3.14 \times 8 \text{ cm} = 25.12$

$C = 25.12$ cm

Find the circumference of a circle with a diameter of 28 ft. Use $\pi = \frac{22}{7}$

$C = \pi \times d$

$d = 28$ ft

$C = \frac{22}{\cancel{7}} \times \frac{\cancel{28}^{4}}{1} = 88$

$C = 88$ ft

PRACTICE

Find the circumference of each circle.
Use $\pi = 3.14$ for Questions 1–3. Use $\pi = \frac{22}{7}$ for Questions 4–6.

1.

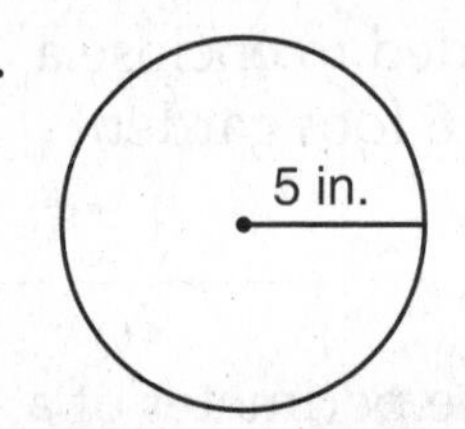

C = ____________

2.

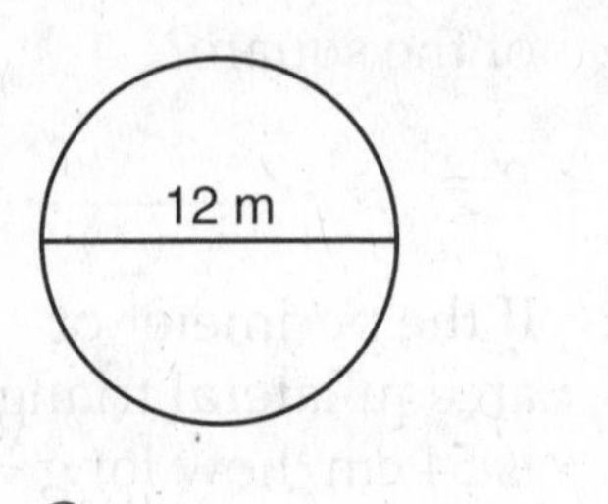

C = ____________

3.

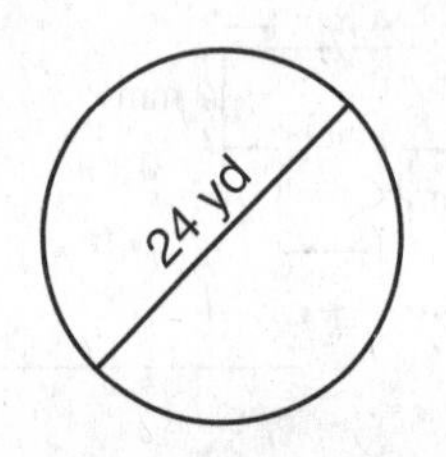

C = ____________

4.

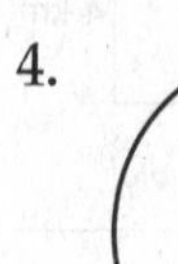

C = ____________

5.

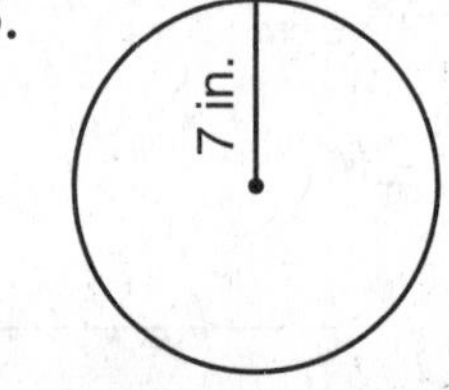

C = ____________

6.

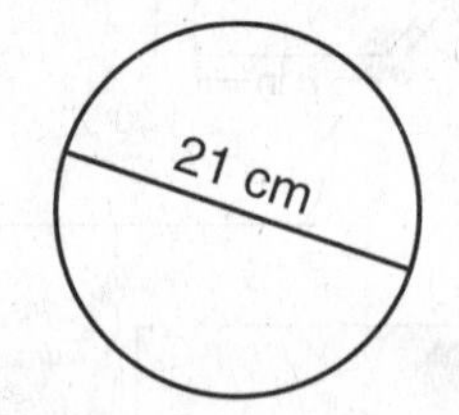

C = ____________

Finding Area of Squares and Rectangles

Area is the measure of the surface within a given region. Area is measured in **square units**. A capital letter *A* is used to represent area.

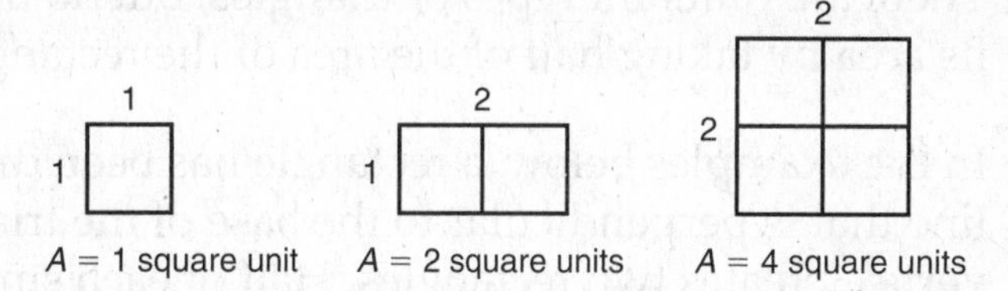

A rectangle with a length of 4 units and a width of 3 units can be divided to form 3 rows with 4 squares in each row, making a total of 12 squares. Each square measures 1 unit on a side. The area of the rectangle is 12 square units, or 12 units^2.

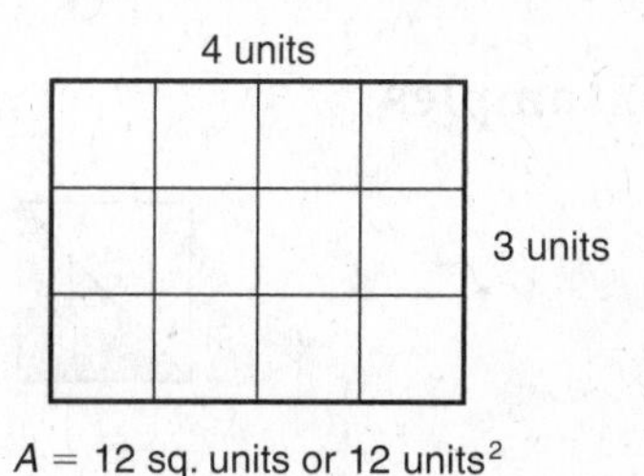

A square with a length of 5 inches on each side can be divided to form 5 rows of 5 squares, or 25 squares that measure 1 inch on each side. The area of the square is 25 square inches, or 25 in.^{2}

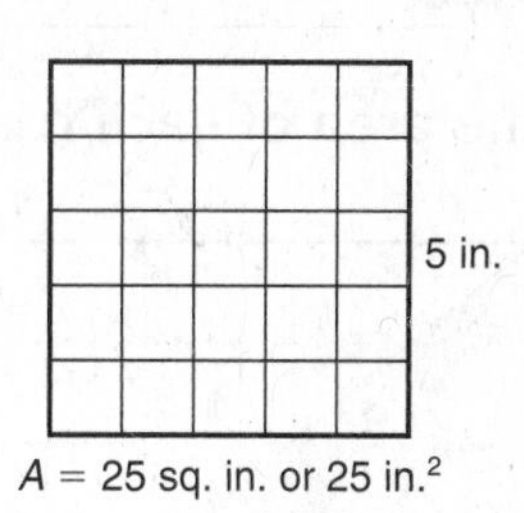

A shortcut for finding the area of a rectangle or square is to multiply the length by the width. You can use the formula $A = lw$, where l represents length, and w represents width.

Example Find the area of a rectangle that measures 8 yd by 10 yd.

- Write the formula. $A = lw$
  length = 10 yd, width = 8 yd
- Replace variables with numbers. $A = 10 \text{ yd} \times 8 \text{ yd}$
- Multiply. $A = 80$ square yards, or 80 yd^2

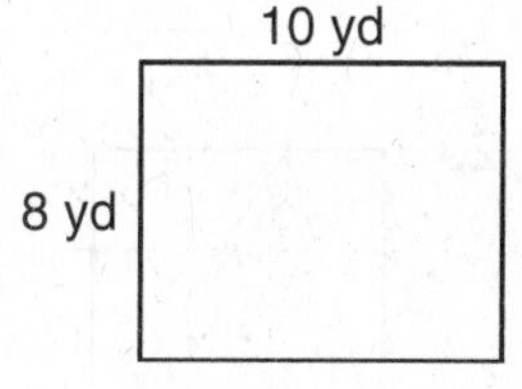

PRACTICE

Find each area. Be sure to label the answer as square units. *Hint*: Divide the figures in Numbers 5 and 6 into two rectangles.

1. 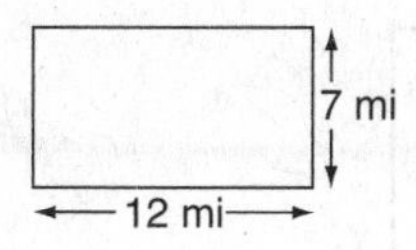

$A =$ ____________

2. a square patio that measures 5 m by 5 m

$A =$ ____________

3. 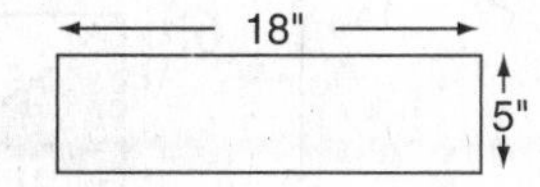

$A =$ ____________

4. a floor that is 12 ft by 15 ft

$A =$ ____________

5. 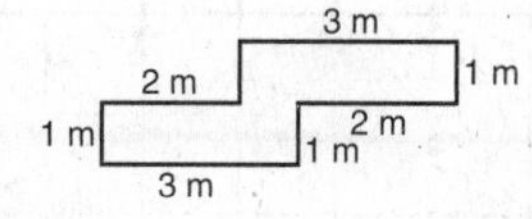

$A =$ ____________

6.

$A =$ ____________

Finding the Area of a Triangle

There are different types of triangles, but no matter what the type of triangle, you can find its area by taking half of the area of the rectangle the triangle "lives in."

In the examples below, a rectangle has been drawn around each triangle. You can see that a line that is perpendicular to the base of the triangle, and that passes through the triangle's vertex, creates two rectangles. Half of each smaller rectangle is shaded, so each triangle equals half of the larger rectangle.

Examples

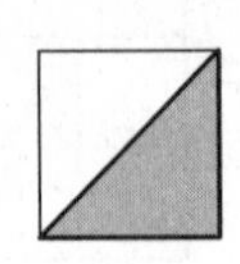

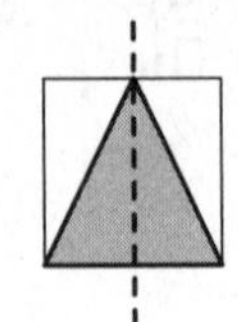

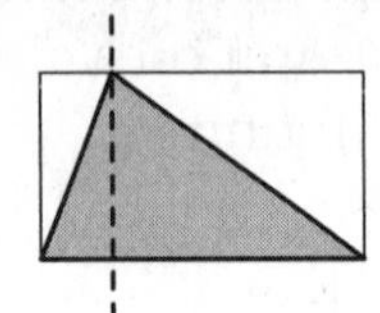

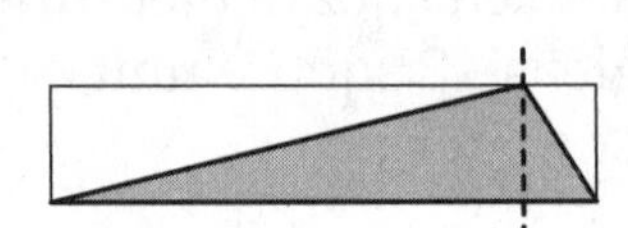

PRACTICE

Find the area of each triangle. Label each answer as square units.

1.

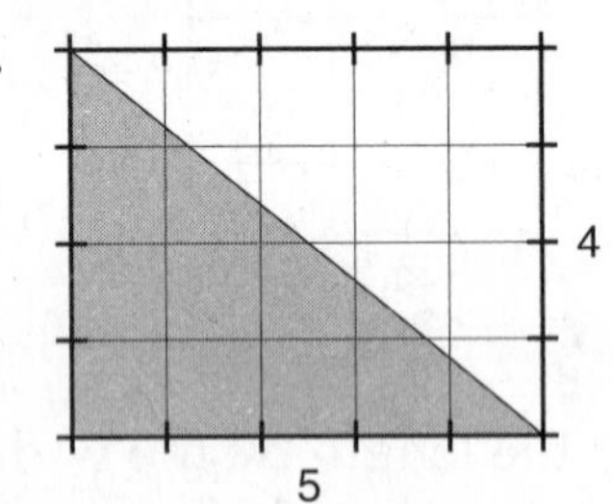

$A =$ ________________

2.

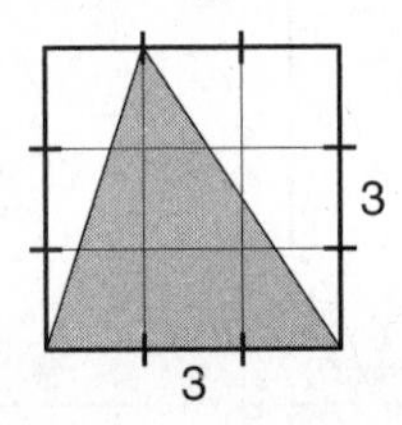

$A =$ ________________

3.

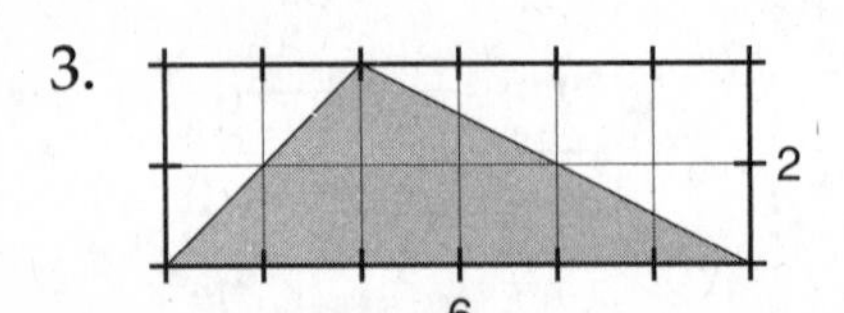

$A =$ ________________

4.

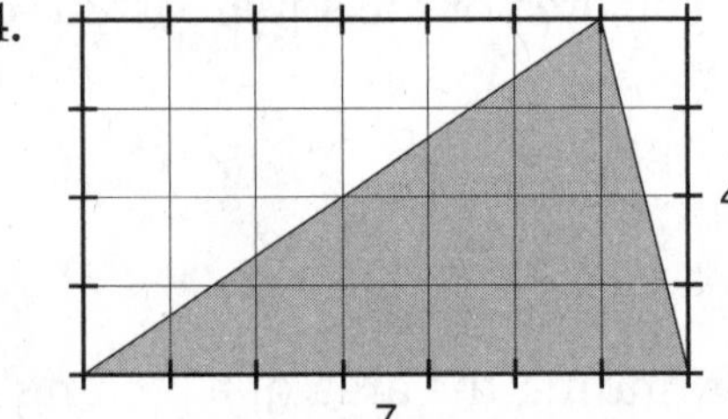

$A =$ ________________

5.

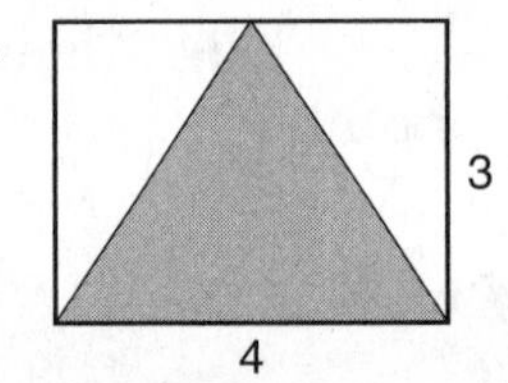

$A =$ ________________

6.

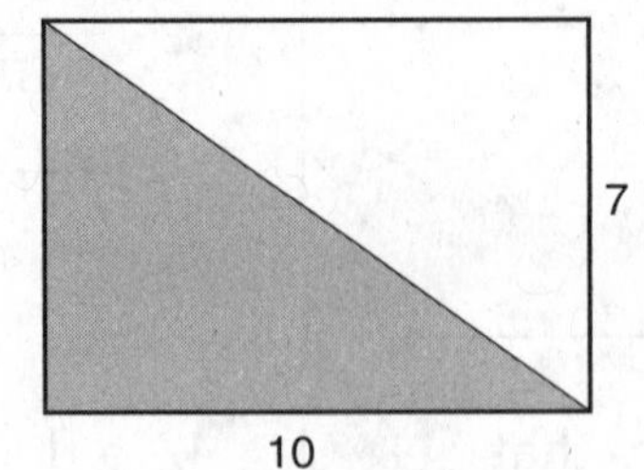

$A =$ ________________

Since a triangle is always half of the rectangle it lives in, you can use a formula to find the area of a triangle. The formula for finding the area of a triangle is $A = \frac{1}{2}bh$, where b represents the base of the triangle, and h represents the height of the triangle.

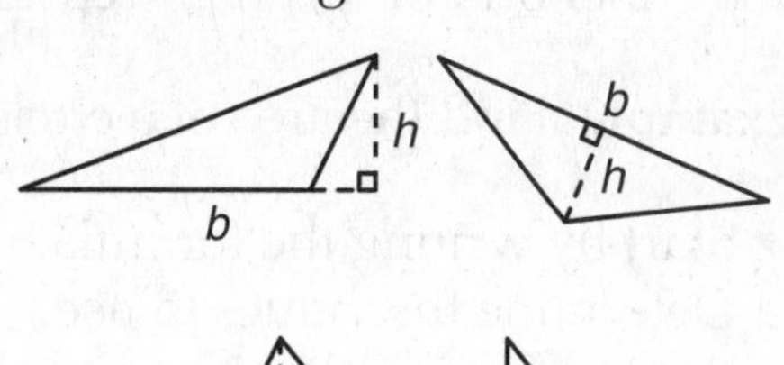

- The base is usually at the bottom of the triangle, but it doesn't have to be.
- The height is the distance between the base and the vertex directly opposite it. This distance must be at a right angle to the base. The height may be shown outside the triangle.

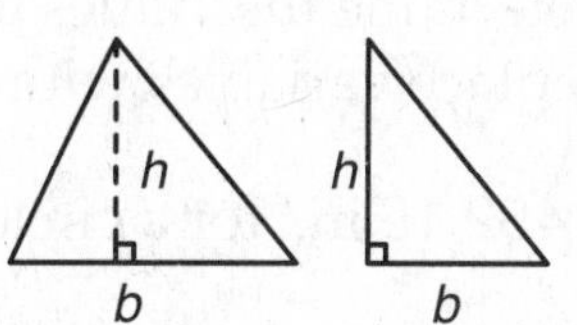

To use the formula, replace b and h with the values they represent in the triangle, and then multiply.

Example Find the area of this triangle.
- Write the formula.
- Replace variables with numbers.
- Multiply.

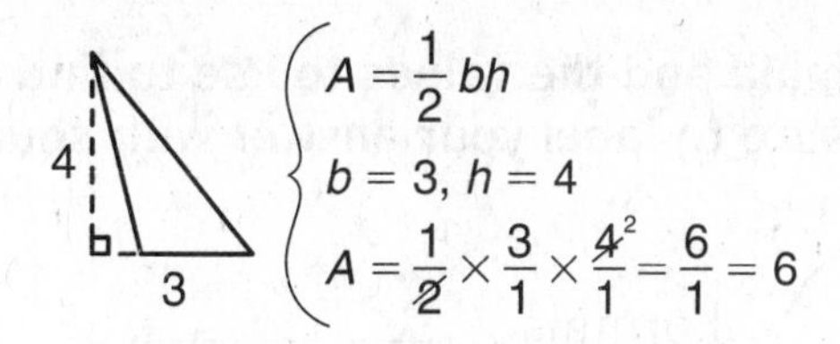

$A = 6$ square units *or* 6 units^2

PRACTICE

Find the area of each triangle. Use the formula $A = \frac{1}{2}bh$. Round your answers to the nearest tenth. Be sure to label your answers with square units.

1.

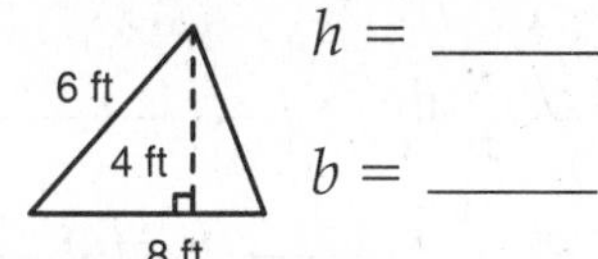

$h =$ ______

$b =$ ______

$A =$ ______

2. 15 m, 16 m, 5 m

$h =$ ______

$b =$ ______

$A =$ ______

3. 20 cm, 24 cm, 24 cm, 24 cm

$h =$ ______

$b =$ ______

$A =$ ______

4.

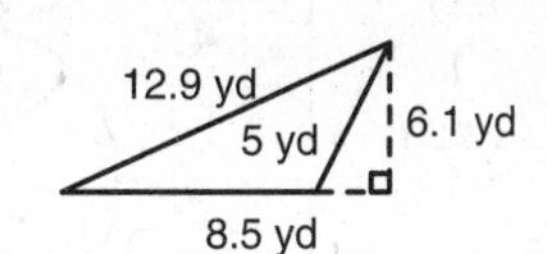

$A =$ ______

5.

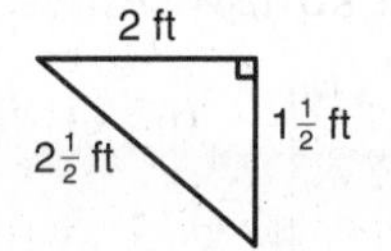

$A =$ ______

6. 5.3 m, 7.5 m, 4.2 m, 8.7 m

$A =$ ______

7. $h = 3\frac{1}{2}$ in., $b = 6\frac{1}{4}$ in.

$A =$ ______

8. $h = 4.5$ m, $b = 7.3$ m

$A =$ ______

9. $h = 4$ cm, $b = 5.35$ cm

$A =$ ______

Finding the Area of a Circle

The area of a circle is the number of square units contained within the circumference of the circle. You can use a formula $A = \pi r^2$ to find the area of a circle, where A stands for area, π is either 3.14 or $\frac{22}{7}$, and r^2 represents the radius squared, which means *radius* × *radius.*

Example Find the area of a circle that has a radius of 12 inches. Use $\pi = 3.14$.

- Start by writing the formula. $A = \pi r^2$
- Determine the values to use. $\pi = 3.14$ $r = 12$ in. $r^2 = 12$ in. × 12 in. = 144 in.^{2}
- Replace variables with values. $A = 3.14 \times 144$ in.^{2} = 452.16 in.^{2}

$A = 452.16$ in.^{2} for a circle with a radius of 12 inches.

PRACTICE

Write the formula and the values to use to find the area of each circle. Use $\pi = 3.14$ to find the area. Be sure to label your answer with square units.

1.

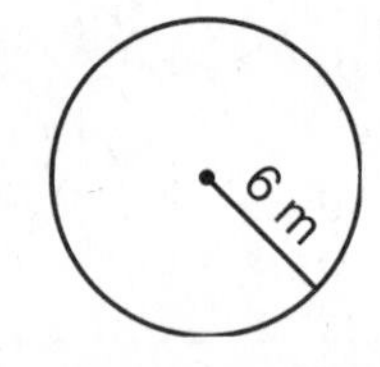

Formula ____________

$r =$ ______ $r^2 =$ ______

$A =$ ____________

2.

Formula ____________

$r =$ ______ $r^2 =$ ______

$A =$ ____________

3.

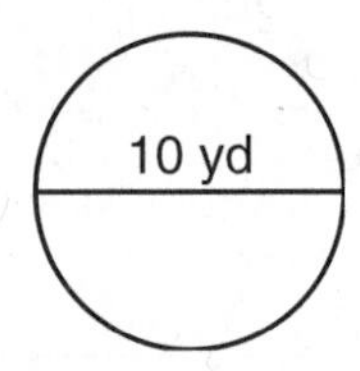

Formula ____________

$r =$ ______ $r^2 =$ ______

$A =$ ____________

4.

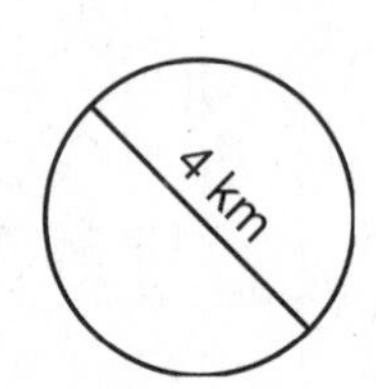

Formula ____________

$r =$ ______ $r^2 =$ ______

$A =$ ____________

Write the formula and the values to use to find the area of each circle. Use $\pi = \frac{22}{7}$ to find the area. Be sure to label your answer with square units.

5. circle with radius of 14 in.

Formula ____________

$r =$ ______ $r^2 =$ ______

$A =$ ____________

6. circle with a diameter of 14 m

Formula ____________

$r =$ ______ $r^2 =$ ______

$A =$ ____________

Finding Surface Area

Squares, rectangles, and triangles are examples of two-dimensional figures. They are flat shapes that have two dimensions; length or height, and width.

The faces of geometric solids such as cubes and pyramids, which are three-dimensional figures, are made up of two-dimensional shapes such as squares, rectangles, and triangles.

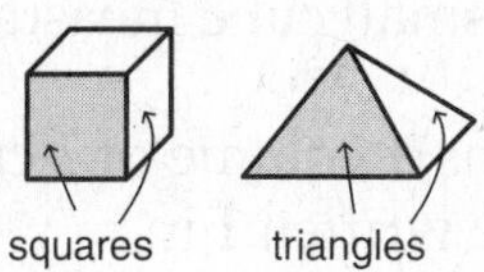

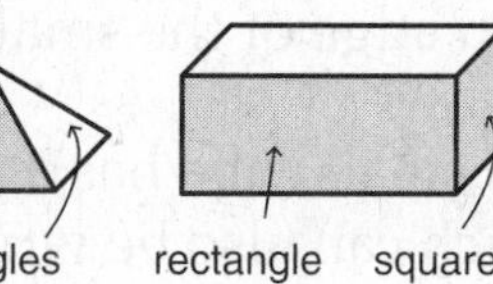

When you find **surface area (SA)**, you find the area of each face of a geometric solid, and then add those areas together to find the total.

Example Find the surface area (*SA*) of the figure. This figure has 6 rectangular faces. Opposite faces are congruent, so you can find the area of one and then double it.

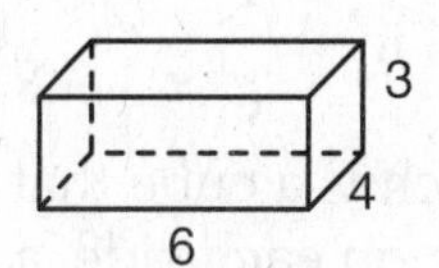

- Find the area of each face.

Top and bottom are each 4 by 6.

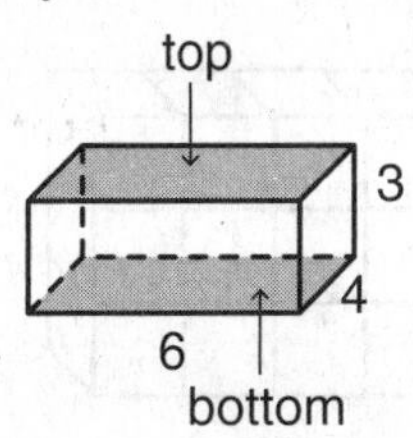

$A = lw = 4 \times 6 = 24$ sq. units
2×24 sq. units $= 48$ sq. units

Front and back are each 3 by 6.

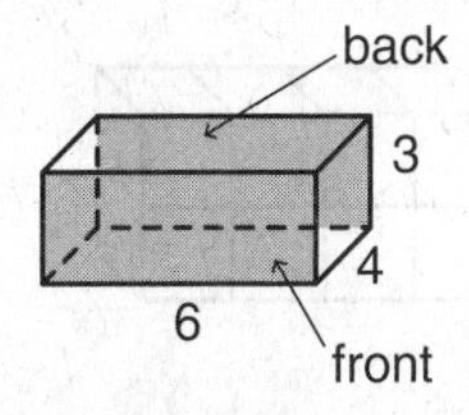

$A = lw = 3 \times 6 = 18$ sq. units
2×18 sq. units $= 36$ sq. units

Right and left sides are each 3 by 4.

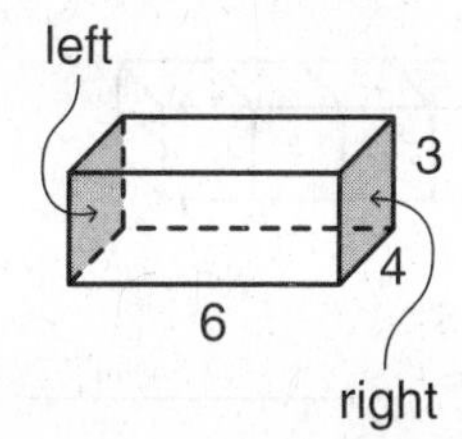

$A = lw = 3 \times 4 = 12$ sq. units
2×12 sq. units $= 24$ sq. units

- Add the areas. $SA = 48 + 36 + 24 = 108$ sq. units

PRACTICE

Find the surface area of each figure.

1.

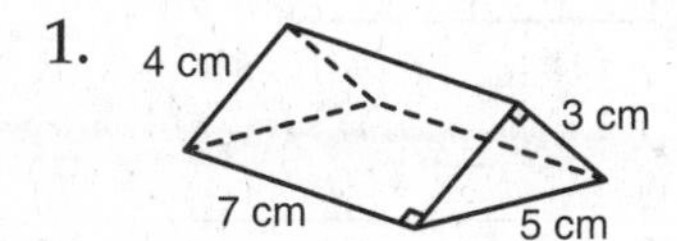

$SA =$ ________

2.

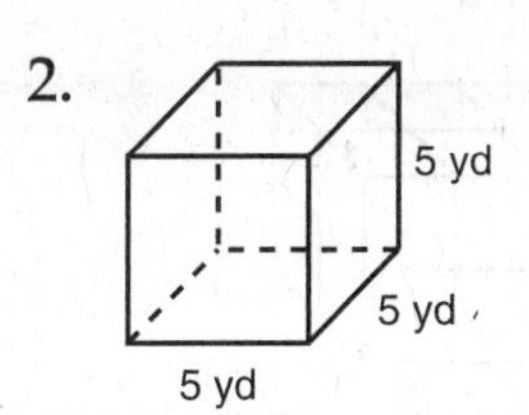

$SA =$ ________

3.

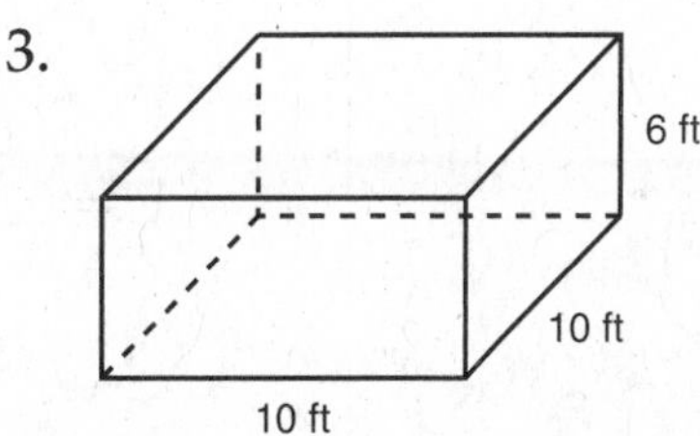

$SA =$ ________

4.

$12\frac{1}{2}$ in.

4 in.

4 in.

$SA =$ ________

Finding Volume

Volume is the space occupied by — or the space inside — a solid, measured in cubes called cubic units. The capital letter V is used to represent volume.

Each edge of the small cube measures 1 inch.

The small cube has a volume of 1 cubic inch, which can also be written 1 $in.^3$

1 in.
1 in.
1 in.

It takes 6 small cubes to make this larger shape, so this larger shape has a volume of 6 cubic inches or 6 $in.^3$

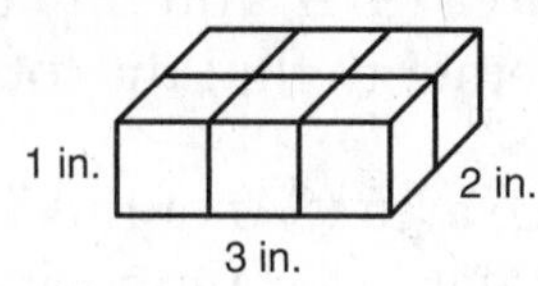

A cubic inch is a cube that is 1 inch long on each side. Similarly, a cubic foot is a cube that is 1 foot long on each side, a cubic meter is a cube that is 1 meter long on each side, and so on.

PRACTICE

Find the volume of each shape if each cube is equal to 1 cm^3.

1. 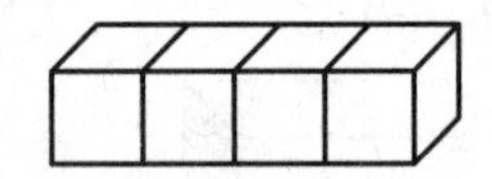

$V =$ ____________

2. 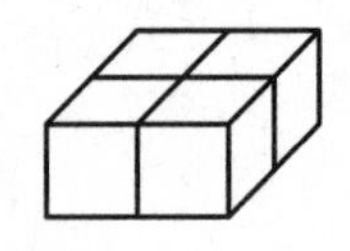

$V =$ ____________

3. 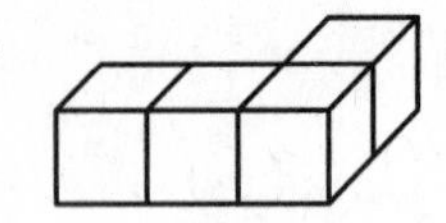

$V =$ ____________

4. 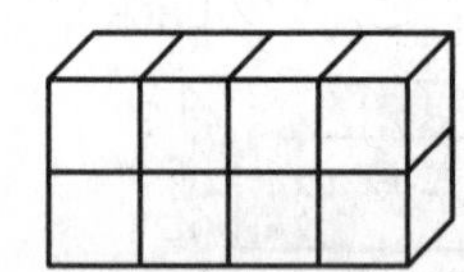

$V =$ ____________

5. 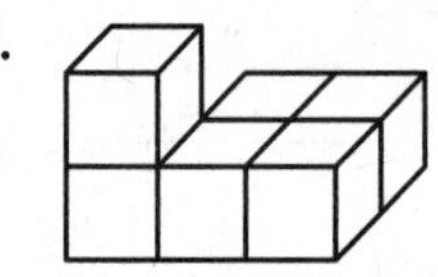

$V =$ ____________

6. 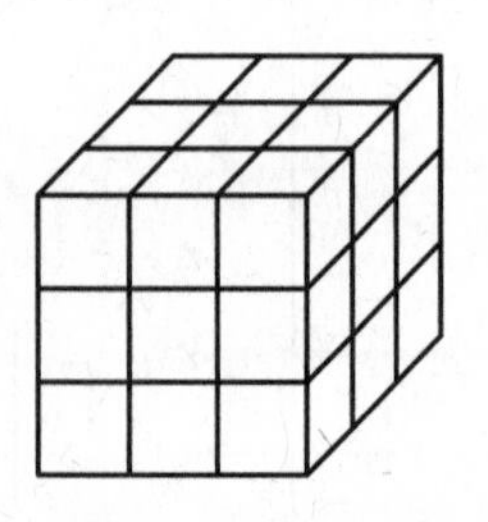

$V =$ ____________

7. 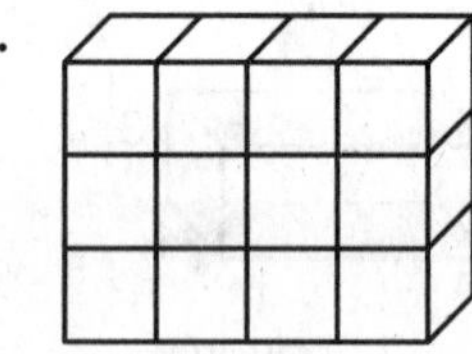

$V =$ ____________

8. 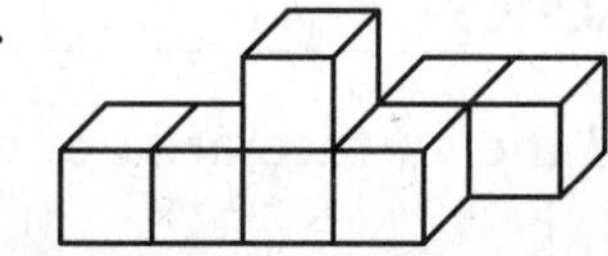

$V =$ ____________

9. 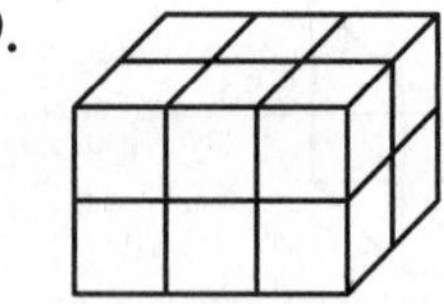

$V =$ ____________

Measurement Skills Checkup

Circle the letter for the correct answer to each question.

This diagram shows the plans for building a bookcase. Study the diagram. Then do Numbers 1–5.

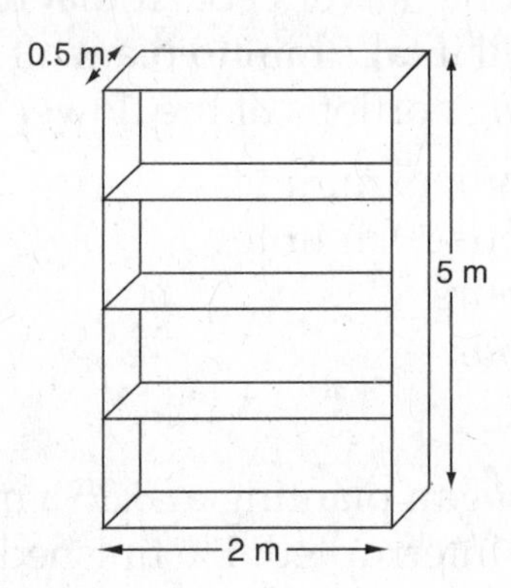

1. Each shelf on the bookcase will be 0.5 (or $\frac{1}{2}$) meter deep. Which of these is another way to describe the depth of each shelf?

   A 5 cm
   B 20 cm
   C 50 cm
   D 95 cm

2. The top of each shelf in the bookcase will be covered with a special adhesive-backed paper. How much of the paper is needed to cover 4 shelves?

   F 1 sq m
   G 4 sq m
   H 8 sq m
   J 10 sq m

3. According to the plans, each shelf of this bookcase can hold 110 kilograms. How many pounds can each shelf hold? (1 kg $\approx$ 2.2 lbs)

   A 5 C 41
   B 242 D 24.2

4. Akira needs $\frac{1}{2}$ gallon of shellac to finish the bookcase, but the shellac comes only in 1-quart cans. How many 1-quart cans does he need?

   F 1 H 5
   G 4 J 2

5. Akira estimates that it will take him $7\frac{1}{2}$ hours to build this bookcase. If he works for 90 minutes a day, how many days will it take him to finish?

   A 4 days C 5 days
   B 6 days D 9 days

Use the figure shown for Numbers 6 and 7.

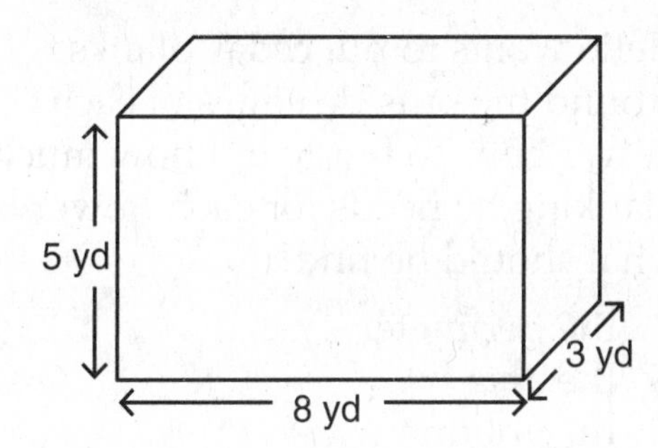

6. What is the volume of the above figure?

   F 120 cubic yd
   G 140 cubic yd
   H 160 cubic yd
   J 180 cubic yd

7. What is the surface area (*SA*) of the figure?

   A 79 sq yd
   B 120 sq yd
   C 136 sq yd
   D 158 sq yd

Measurement Skills Checkup (continued)

This diagram shows plans for two flower beds that Helio is going to put along the sides of his driveway. Study the diagram. Then do Numbers 8–12.

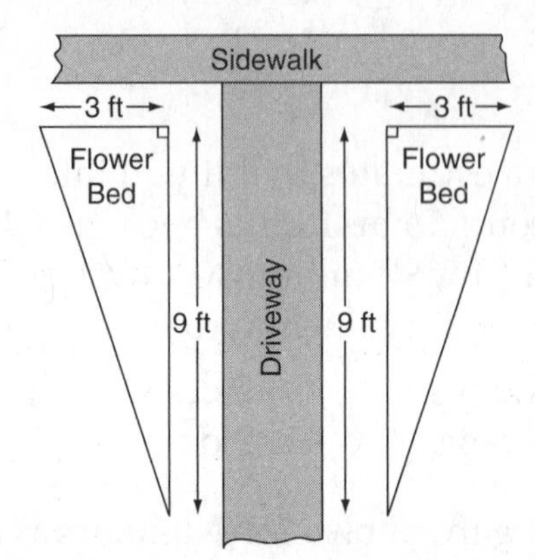

8. Helio wants to put cedar planks around the outside edges of each flower bed. To figure out how much planking he needs for each flower bed, what should he find?

   F the perimeter
   G the area
   H the volume
   J the depth

9. Helio uses an ounce of fertilizer for each square foot of area. How much fertilizer does he need to cover both flower beds?

   A 18 ounces **C** 27 ounces
   B 54 ounces **D** 108 ounces

10. Helio buys $\frac{1}{4}$ of a pound of fertilizer. How many ounces is that?

    F 25 **H** 5
    G 4 **J** 8

11. It takes 20 minutes for Helio to plant half of one flower bed. At that rate, how long will it take him to plant all the *remaining* portions of the flower beds?

    A 1 hour 15 minutes
    B 1 hour 20 minutes
    C 2 hours
    D 1 hour

12. Helio began planting at 8:25 a.m. but he was interrupted. He finished at 2:15 p.m. How much time elapsed between his starting and finishing times?

    F 5 hours 50 minutes
    G 6 hours 50 minutes
    H 5 hours 90 minutes
    J 5 hours 10 minutes

13. What is the circumference of a circle with a radius of 4 ft? Use $\pi = 3.14$.

    A 12.56 ft **C** 25.12 ft
    B 50.24 ft **D** 66.24 ft

14. Which is the best estimate of the capacity of a bathtub?

    F 40 milliliters
    G 40 centiliters
    H 40 liters
    J 40 kiloliters

15. Which measurement is equivalent to $3\frac{1}{2}$ yards?

    A 12 ft **C** 10 ft
    B $10\frac{1}{2}$ ft **D** $9\frac{1}{2}$ ft

Probability, Data, and Statistics

Investigating Probability

Probability is the study of how likely it is for an **event** to occur. Possible results are called **outcomes**. Probability can be expressed as a number from 0 to 1. You can find this number by comparing the number of ways that are favorable for the event to occur to all of the outcomes that are possible.

$$\text{Probability} = \frac{\text{number of favorable outcomes}}{\text{total number of outcomes possible}}$$

- If there is **no chance** that an event can happen, the probability is 0. For example, the probability of a banana turning into an apple would be 0.
- If an event is **certain** to happen, its probability is 1, or 100%. For example, the probability that you will get wet if you stand in the rain without a raincoat or an umbrella is 1, or 100%.
- The greater the probability that an event will occur, the closer the number representing the probability will be to 1; the less the probability, the closer the number will be to 0.

When you toss a coin, the probability of tossing heads is $\frac{1}{2}$. This is because there are 2 possible outcomes, heads or tails. Only 1 of those outcomes is favorable for heads.

$$\text{Probability (heads)} = \frac{\text{favorable outcomes}}{\text{possible outcomes}} = \frac{1}{2}$$

The probability for heads on a single toss can be expressed in any of the following ways:

fraction	**decimal**	**ratio**	**percent**	**words**
$\frac{1}{2}$	0.5	1:2	50%	The chances are one in two. There is a *fifty percent* or a *fifty-fifty* chance.

Example

Art, Etan, José, Ryan, and Dwight wrote their names on separate slips of paper and put them in a box. One name will be drawn to win a prize. Each person has an equal chance of winning. What is the probability that Ryan's name will be drawn?

There are 5 different names that can be drawn. Only 1 name is Ryan's.

$$\text{Probability (of Ryan's name being drawn)} = \frac{\text{favorable outcomes}}{\text{possible outcomes}} = \frac{1}{5} = 0.2 = 20\%$$

Ryan has a 1 in 5, or a 0.2, or a 20% chance of having his name drawn.

Example

There are 12 towels in a laundry bag. Three are green, 4 are white, and 5 are yellow. If Tammy reaches into the bag without looking and pulls 1 towel out, what is the probability that she will select a green towel?

$$\text{Probability} = \frac{\text{number of favorable outcomes}}{\text{total number of outcomes}} = \frac{\text{3 green towels}}{\text{12 towels}} = \frac{3}{12} = \frac{1}{4} = 0.25 = 25\%$$

The probability that Tammy will select a green towel is $\frac{1}{4}$, or 0.25, or 25%.

Investigating Probability (continued)

PRACTICE

Circle the correct answer.

1. A box contains 3 green and 5 red counters. What is the probability of reaching into the box without looking and picking one red counter?

   A $\frac{1}{3}$ C $\frac{3}{5}$

   B $\frac{3}{8}$ D $\frac{5}{8}$

2. What is the probability that if someone picks a number from 1 through 10, the number will be even?

   F $\frac{1}{5}$ H $\frac{1}{2}$

   G $\frac{2}{3}$ J $\frac{1}{4}$

3. In the class election, Carlo got 12 votes, Bryan got 14 votes, and Roger got 10 votes. If a person in the class is selected at random, what is the probability that he or she voted for Roger?

   A $\frac{1}{3}$ C $\frac{1}{10}$

   B $\frac{5}{18}$ D $\frac{5}{13}$

4. What is the probability that a person picked at random is born in a month beginning with the letter "J"?

   F $\frac{1}{12}$ H $\frac{1}{3}$

   G $\frac{1}{4}$ J $\frac{1}{6}$

Use the diagram of lettered discs for Numbers 7 and 8. There are 2 discs labeled C, 3 labeled B, and 5 labeled A.

5. If you reach into the bag without looking and take out a disc, what is the probability it will be a "C"?

   A 25% C 20%
   B 10% D 2%

6. What is the probability you will pick the letter "D"?

   F 100% H 25%
   G 50% J 0%

7. What is the probability that if you drop something, it will fall down, not up?

   A $\frac{1}{2}$ C 0.25

   B 3:4 D 100%

8. What is the probability that a letter picked at random from the first 10 letters of the alphabet will be a vowel?

   F $\frac{3}{10}$ H $\frac{2}{5}$

   G $\frac{4}{7}$ J $\frac{3}{5}$

Gathering Data for a Survey

When a company or an organization wants to know what people think, they conduct a **survey**. Information is usually gathered from a **sample group**, which is a small part of the **population** they want to know about.

- The population consists of people or objects that the survey is about. For example, in a survey to find out how many people between the ages of 20 and 30 listen to a certain radio station, the population is all of the people between 20 and 30 who live in the listening area of the station.
- The sample group selected to be surveyed should be members of the population.
- The sample group should represent a **random sample**. This means that every member of the population has an equal chance of being selected. The survey of radio listeners should include people 20 to 30 years old from all walks of life who live throughout the entire listening area rather than selecting from only one group or one neighborhood.

It is important that survey questions not be biased; questions should not influence the answers. Look at these two questions designed to find what kind of food people like:

What kind of pizza do you like?
What is your favorite food?

The first question assumes a person would select pizza as a favorite food, while the second question leaves the choice open to any type of food.

PRACTICE

Circle the correct answer.

1. Jamie wants to expand the cookware section in her gift shop. Which of these sample groups would give her the best data about what items she should add?
   A people who buy greeting cards at her store
   B people who take cooking classes at a local cooking school
   C a random sample of people who have purchased only cookware items in her shop
   D a random sample of customers to her shop

2. Herman is writing a questionnaire for a survey to find out whether people recycle newspapers. Which of these questions would give him the most unbiased results?
   F Are you a person who cares about the environment and recycles newspapers?
   G Where do you take your newspapers to recycle them?
   H What types of items do you take to the recycling center?
   J Do you recycle newspapers?

3. Which question would get the most unbiased results about a person's reaction to a movie?
   A Weren't the costumes great?
   B What was your favorite part of the movie?
   C What did you think of the movie?
   D Did you think this movie was as good as the last one starring the same actor?

Reading a Table

Tables and graphs are useful ways to organize data. They allow us to read information quickly and easily. To understand a table or a graph, always begin by reading the title and the headings. They explain the relationships shown in a table or graph.

Top 10 Network Telecasts of All Time

Rank	Program	Telecast Date	Household Rating*	Number of Household Surveyed
1	M*A*S*H Special	Feb. 28, 1983	60.2%	50,150,000
2	Dallas (Who Shot J.R.?)	Nov. 21, 1980	53.3%	41,470,000
3	Roots, Part VIII	Jan. 30, 1977	51.1%	36,380,000
4	Super Bowl XVI	Jan. 24, 1982	49.1%	40,020,000
5	Super Bowl XVII	Jan. 30, 1983	48.6%	40,480,000

*Percentage of U.S. households surveyed that viewed the show
Source: Courtesy Nielsen Media Research

PRACTICE

Use the table above to answer each question.

1. This column of numbers is taken from the table. What do these numbers show?

60.2%
53.3%
51.1%
49.1%
48.6%

A the percentage of Americans who say each show is their favorite
B the percentage of households surveyed that viewed the show
C the percentage of judges who voted for each show

2. On what date was the "Who Shot J.R.?" episode of Dallas shown? ________

3. How many households viewed the last episode of M*A*S*H when it was shown in February of 1983? ________

4. Which shows had a viewer rating higher than 50%? ________

5. Which show had a viewer rating of 60.2%? ________

6. What is the earliest date listed in this table? ________

7. For which show were the greatest number of households surveyed? ________

Using a Price List

A menu or price list is a common type of table. To use a menu or price list, find what you want to buy. Then look for the price for that item. Do not get confused if you are looking for several prices. Write the price for each item. Then do any figuring.

Parcel Post Rates

Weight	Zones 1 and 2	Zone 3	Zone 4	Zone 5	Zone 6	Zone 7	Zone 8
1 pound	$3.69	$3.75	$3.75	$3.75	$3.75	$3.75	$3.75
2 pounds	3.85	3.85	4.14	4.14	4.49	4.49	4.49
3 pounds	4.65	4.65	5.55	5.65	5.71	5.77	6.32
4 pounds	4.86	5.20	6.29	6.93	7.14	7.20	7.87
5 pounds	5.03	5.71	6.94	7.75	8.58	8.64	9.43
6 pounds	5.63	6.01	7.44	8.50	9.52	9.90	11.49
7 pounds	5.80	6.28	7.91	9.20	10.35	11.39	12.83
8 pounds	5.98	6.53	8.30	9.84	11.11	12.54	15.04
9 pounds	6.11	6.76	8.74	10.45	11.83	13.38	17.04
10 pounds	6.28	7.57	9.10	11.01	12.50	14.17	18.14

Insurance rates: $1.30 to insure packages for up to $50; $2.20 to insure packages for $50.01–$100; $3.20 to insure packages for $100.01–$200; $4.20 to insure packages for $200.01–$300.
Source: U.S. Post Office

PRACTICE

Use the price list above to answer the following questions.

1. How much does it cost to mail a 3-pound package to Zone 3? ________

2. How much more does it cost to send a 4-pound package to Zone 6 than to send it to Zone 2? ________

3. What is the total cost to send a 2-pound package to an address in Zone 1, and an 8-pound package to Zone 3? ________

4. How much would it cost to send a 9-pound package to Zone 4 and insure it for $250? ________

5. To the nearest dollar, how much would it cost to send a 2-pound package to Zone 8, a 4-pound package to Zone 4, and a 5-pound package to Zone 6? ________

6. You have $5 to send an 3-pound package to Zone 1. How much insurance can you afford? ________

7. You are sending a package to Zone 6. How much more would it cost to send a 9-pound package rather than a 5-pound package? ________

Using Tables to Make Comparisions

You can use the numbers in a table to make comparisons.

- You can find the **difference** between two numbers by subtracting.
- You can see how **many times larger** one number is than another by setting up a ratio.
- You can find **what fraction** one number is of another by forming a ratio.

The 10 Largest Native American Nations in the United States

Rank	Nation	Population	Percent
1	Cherokee	308,132	16.4
2	Navajo	219,198	11.7
3	Chippewa	103,826	5.5
4	Sioux	103,255	5.5
5	Choctaw	82,299	4.4
6	Pueblo	52,939	2.8
7	Apache	50,051	2.7
8	Iroquois	49,038	2.6
9	Lumbee	48,444	2.6
10	Creek	43,550	2.6

Source: U.S. Department of Commerce 1990 Census

Example The Chippewa Nation is about how many times larger than the Lumbee?

Round the two numbers and then set up a ratio to compare the two nations. Put the greater number for the Chippewa in the numerator.

$$\frac{\text{Chippewa}}{\text{Lumbee}} \quad \frac{100{,}000}{50{,}000} = \frac{2}{1} = 2 \text{ times}$$

The Chippewa Nation is 2 times larger than the Lumbee.

Example The Creek Nation is approximately what fraction of the size of the Chippewa?

Round the numbers and put them in fraction form. For this comparison, put the greater number for the Chippewa in the denominator.

$$\frac{\text{Creek}}{\text{Chippewa}} \quad \frac{40{,}000}{100{,}000} = \frac{2}{5}$$

The Creek Nation is about $\frac{2}{5}$ the size of the Chippewa.

PRACTICE

Use the table above to fill in the blanks.

1. What is the largest Native American group in the United States? ______
2. Which tribe is larger, the Apache or the Choctaw? ______
3. How many more Lumbee are there than Creek? ______
4. Which group is closest in size to the Chippewa? ______
5. Which group is about $\frac{2}{3}$ the size of the Cherokee? ______
6. How many more Choctaw are there than Pueblo? ______
7. Together, the Apache and the Pueblo are about the same size as which group? ______
8. The Sioux Nation is about what fraction of the size of the Navajo?

   A $\frac{1}{5}$ **C** $\frac{1}{4}$

   B $\frac{1}{3}$ **D** $\frac{1}{2}$

Interpreting Information in Tables

Tables are one way to display data that have been collected. The table below shows the results of a survey of about 25,196,000 veterans, taken to find out how satisfied they were with their ability to get information about their benefits.

Veterans' Satisfaction on Accessing Benefits Information

Very satisfied	15.4 %
Somewhat satisfied	30.4 %
Neither satisfied nor dissatisfied	24.9 %
Somewhat dissatisfied	13.6 %
Very dissatisfied	9.2 %
Don't need information	2.0 %
Unknown	4.5 %

Source: Department of Veterans Affairs

PRACTICE

Use the above table to answer these questions.

1. What is the approximate ratio of the percent of veterans who were somewhat satisfied to those who were very satisfied with their ability to get benefits information?

   A 2 to 1 **C** 1 to 4
   B 1 to 2 **D** 4 to 1

2. About how many times as many veterans were either somewhat or very satisfied compared with those who were somewhat or very dissatisfied with their ability to get benefits information?

   F half as many
   G two-thirds as many
   H twice as many
   J three times as many

3. Approximately what fraction of the veterans surveyed reported that they were neither satisfied nor dissatisfied?

   A $\frac{1}{3}$ **C** $\frac{1}{4}$

   B $\frac{1}{5}$ **D** $\frac{1}{2}$

4. About how many veterans reported they did not need information on benefits?

   F $\frac{1}{4}$ million **H** $\frac{1}{2}$ million

   G 1 million **J** 2 million

5. Circle the letter of the graph that most accurately displays the data given in the table above.

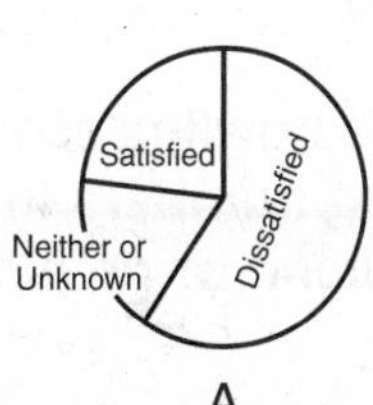

A

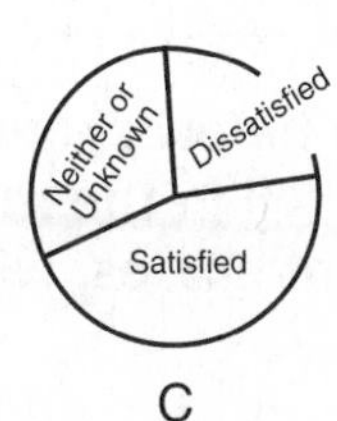

C

Dissatisfied
Satisfied
Neither or Unknown
B

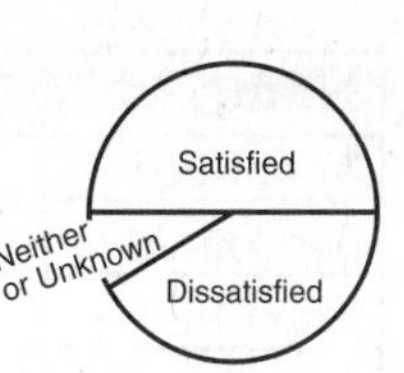

D

Finding Mean, Median, and Mode

Often, when a set of data is collected, the numbers in the data are used to get a single number that represents the data in one way or another.

Example Here are Rob's scores for the last 10 times he bowled.
151, 167, 139, 152, 163, 169, 201, 171, 168, 169
Find the range of Rob's bowling scores.
Find the mean, median, and mode of Rob's scores.

- **Minimum** The minimum is the least number in the set of data.
  Rob's minimum score is 139.
- **Maximum** The maximum is the greatest number in the set of data.
  Rob's maximum score is 201.
- **Range** The difference between the maximum and minimum is the range.
  For Rob's scores, $201 - 139 = 62$. Rob's range is 62.
- **Mean or average** The mean is the result of distributing the data to make each of the numbers the same. To find the mean, add all of the numbers, and then divide the sum by the number of amounts you added.
  Rob's 10 scores add to 1,650. $1{,}650 \div 10 = 165$ Rob's average is 165.
- **Median** The median is the number in the middle of the list after the numbers have been arranged in order from least to greatest. If the number of values that were added is even, there will be two middle numbers; find the average of those two numbers.
  139, 151, 152, 163, 167, 168, 169, 169, 171, 201
  There are two middle scores. Find their average.
  $167 + 168 = 335$ $335 \div 2 = 167\frac{1}{2}$ Rob's median score is $167\frac{1}{2}$.
- **Mode** The mode is the value that appears most often.
  There can be more than one mode. If no number appears more often than any other, the set of data has *no mode*.
  Rob scored 169 twice. The mode of Rob's scores is 169.

PRACTICE

The table below shows the bowling scores of Rob's teammates. Each player bowled 4, 5, or 6 games. Fill in the missing numbers in the shaded section of the table. (If there is no mode for a set of scores, write *none* or *no mode*.)

Bowling Scores

Name	1	2	3	4	5	6	Median	Mode	Mean
Matt Reed	175	180	196	200	179	154	179.5	no mode	181
Ben Hanks	189	156	168	207	180	—			
Jose Ruiz	232	230	263	180	230	—			
Phil Chu	178	167	172	175	175	165			
Linda Glass	167	163	167	163	—	—		no mode	
Tim Horne	144	127	85	144	116	110			
Connie Chu	212	156	164	246	260	246			

Reading a Circle Graph

Graphs are drawings or diagrams that allow you to read information quickly. You will see various types of graphs in newspapers and magazines, as well as on television.

One type of graph is the **circle graph**. The circle graph, often called a pie graph, divides a circle into wedges that show parts of a whole. Each wedge or section represents a fraction of the total—the larger the section, the greater the fraction of the whole represented.

In a circle graph, it is easy to compare each section with the whole. It is also easy to compare one section with another.

PRACTICE

(*According to Country of Last Residence)
Source: U.S. Census Bureau

Use the graph above for Numbers 1–3. Circle the correct answer.

1. The greatest number of people visiting the United States came from

   A Asia **C** Europe
   B North America **D** South America

2. A little more than $\frac{1}{4}$ of the people who visited the United States were from this continent.

   F Asia **H** Europe
   G North America **J** South America

3. The ratio of visitors from Europe to the visitors from South America was closest to

   A 1 to 2 **C** 1 to 3
   B 3 to 1 **D** 4 to 1

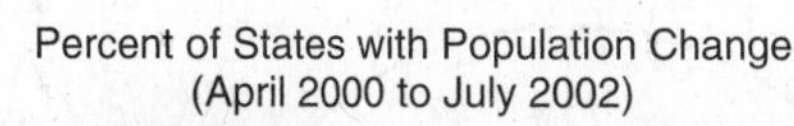

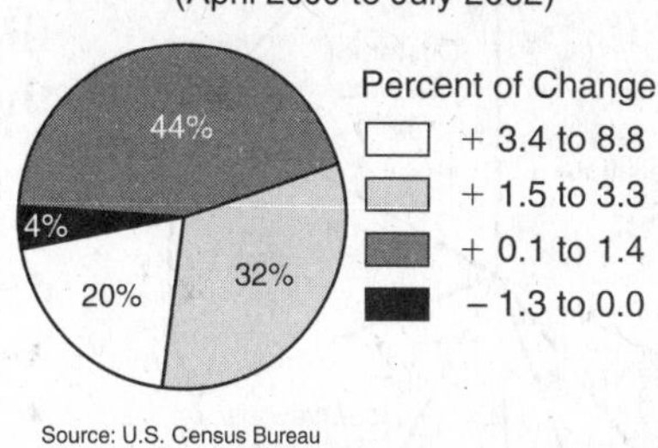

Source: U.S. Census Bureau

The graph above shows the percent of states according to the percent of change in population. Use the graph to answer 4–6.

4. The greatest number of states had a change in population of

   A + 1.5 to 3.3 percent
   B + 0.1 to 1.4 percent
   C + 3.4 to 8.8 percent
   D 0 to −1.3 percent

5. Approximately half of the states had a population change of

   F 3.4 to 8.8 percent
   G 1.5 to 3.3 percent
   H 1.5 to 8.8 percent
   J 0.1 to 3.3 percent

6. A little over three-fourths of the states had a population change of

   A + 0.1 to 3.3 percent
   B + 1.5 to 3.3 percent
   C + 0.1 to 1.4 percent
   D + 3.4 to 8.8 percent

Finding the Numbers Represented in a Circle Graph

Below is a graph of different types of complaints made against top U.S. airlines from 1990-1999. During that 10-year period, a total of 71,245 complaints were registered. Each category has been rounded to the nearest whole percent.

While you can see which category got the greatest percentage of the complaints, this graph does not tell how many complaints there were. To find a number for the complaints for any category, multiply the total number of complaints received by the percent that were received by the category. Since the percents have been rounded, the number of complaints will not be exact.

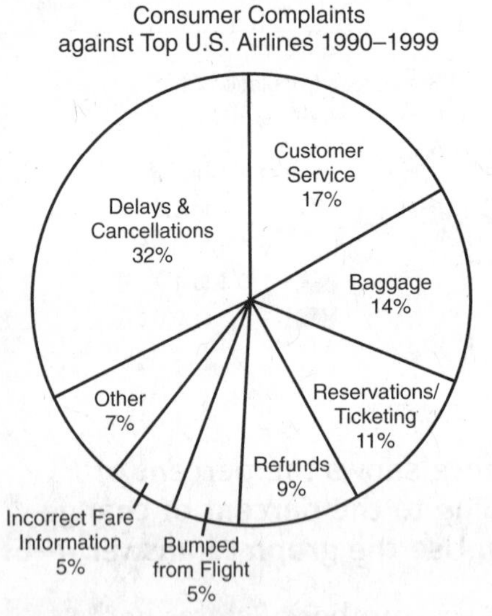

Note: This graph represents a total of 71, 245 complaints.
Source: U.S. Department of Transportation

Example

Approximately how many complaints were there about flights that were late or canceled? Round your answer to the nearest thousand.

- A total of 71,245 complaints were received.
- 32% of the complaints received were for delayed or canceled flights.
- Find 32% of 71,245.

$$32\% \text{ of } 71{,}245 = 0.32 \times 71{,}245 = 22{,}798.4$$
$$= 22{,}798.4 \approx 23{,}000$$

There were about 23,000 complaints about delayed or canceled flights.

PRACTICE

Use the graph above to answer the following questions. Round each answer to the nearest thousand.

1. Approximately how many customer service complaints were received? ________
2. Approximately how many complaints were received regarding baggage? ________
3. About how many complaints were about incorrect fare information? ________
4. About what number of complaints involved problems with reservations or ticketing? ________
5. Which two categories together made up almost half of the complaints received? ________
6. Approximately how many more complaints were received about problems with baggage than about reservation and ticketing problems? ________

Reading a Bar Graph

A bar graph uses thick lines, or bars, to represent categories. Either horizontal or vertical bars can be used. Categories for the bars are listed along one **axis**, and numbers are listed along the other axis.

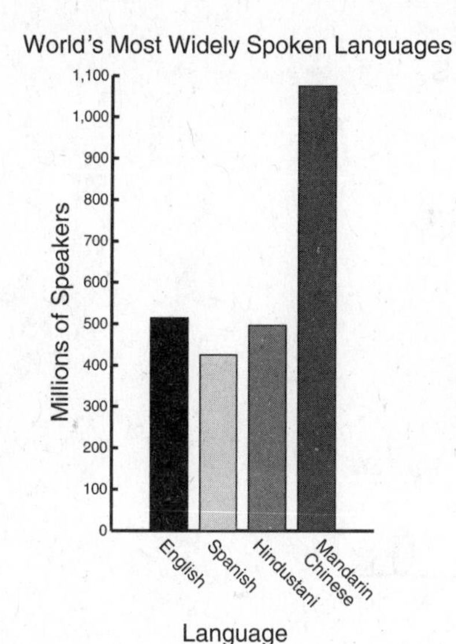

To read a bar graph, identify the bar representing the category you want. Then determine the number that corresponds with the end of the bar. If a bar ends between two numbers, estimate the number for the end of the bar.

Example

About how many people speak Hindustani?

The bar representing Hindustani lies about $\frac{9}{10}$ of the distance between 400 million and 500 million, or about $\frac{9}{10}$ of 100 million. $\frac{9}{10}$ of 100 million is 90 million.

Add 90 million to 400 million to get 490 million.

About 490 million people speak Hindustani.

PRACTICE

Use the graph above to answer the following questions.

1. About how many people in the world speak English? __________

2. Which of these is the best estimate of how many people speak Mandarin Chinese?

   A 1,000
   B 1 billion (1,000,000,000 or 1,000 million)
   C 1.2 billion
   D 1.02 billion

3. About how many times as many people speak Mandarin as speak English? __________

4. Which of the following is *not* shown on this graph?

   A the language that is spoken by the greatest number of people
   B how many more people speak Hindustani than Spanish
   C how many people learn Mandarin as their first language

Understanding Data in Bar Graphs

Bar graphs can be used to show information about more than one set of data at a time. In such graphs, the bars for each set look different. A **key** or **legend** explains what the different bars represent. In the graph below, the key indicates that the different bars represent data for 1960, 1980, and 2001.

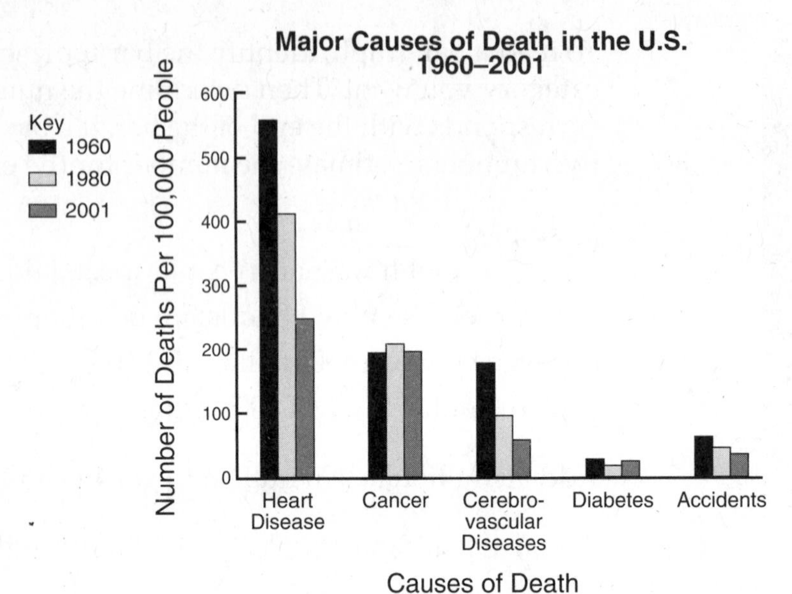

PRACTICE

Use the graph above to answer the following questions.

1. In 2001, about how many Americans out of every 100,000 died of cancer? ________

2. About how many Americans died of heart disease in 1960? ________

3. Which cause of death rose between 1960 and 1980? ________

4. Which two causes of death have had the least change in number over the years shown? ________

5. Which cause of death was about the same in 1960 and 2001? ________

6. About how much lower was the cause of death due to heart disease in 2001 than it was in 1960? ________

7. Based on the graph, an average American town of 200,000 might have expected about how many cancer deaths in 2001? ________

8. Which cause was responsible for around half as many deaths in 2001 as in 1980? ________

9. Which cause of death shows the greatest decrease from 1960 to 2001? ________

Reading a Line Graph

In a **line graph**, line segments connect points or dots that represent data. Line graphs show how something changes over a period of time. When the line displays a pattern of continuing increase, the graph is said to show an **upward trend**. If there is a continuing decrease, the graph shows a **downward trend**. If the data show neither an upward nor a downward trend, the graph is said to show **no trend**.

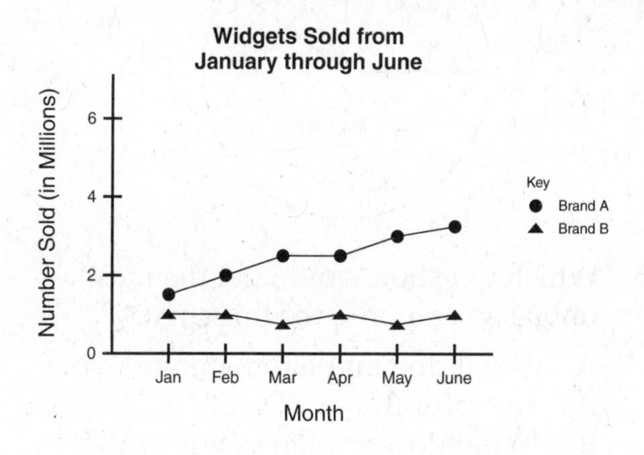

As with any graph, read the titles to find out what information is represented by the graph. Check the legend to identify the symbols used. The graph shown here compares the number of widgets sold by two companies each month for six months.

PRACTICE

Use the above graph to answer the following questions. Circle the correct answer.

1. The graph shows that

   A more Brand A than Brand B widgets were sold during the time shown

   B more Brand B than Brand A widgets were sold during the time shown

   C Brand A and Brand B sold the same number of widgets

2. The sale of Brand A widgets shows

   F an upward trend

   G a downward trend

   H no trend

3. During which month did one brand have about twice as many sales as the other?

   A January　　C February

   B May　　D June

4. In April, approximately how many fewer Brand B widgets were sold than Brand A?

   F $\frac{1}{2}$ million　　H 1 million

   G $1\frac{1}{2}$ million　　J 2 million

5. If sales continue in the same way, approximately how many widget sales can Brand B expect for the month of August?

   A about 4 million

   B about 3 million

   C about 2 million

   D about 1 million

6. If sales continue in the same way, about how many widget sales can Brand A expect for the month of July?

   F $\frac{1}{2}$ million　　H 1 million

   G 3 million　　J 4 million

Probability, Data, and Statistics Skills Checkup

Circle the correct answer for each question.

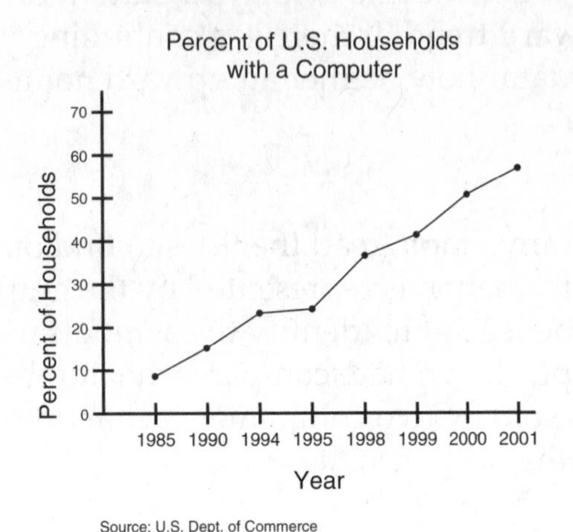

Use the above graph for Numbers 1–4.

1. Which of these is the best estimate of how many more households had a computer in 2000 than in 1985?

   A 5% more C 15% more
   B 40% more D 80% more

2. If in 1995 there were about 100 million households in the United States, about how many households had computers?

   F 24 million H 2.4 million
   G 240 million J There is no way to tell.

3. Which of the following is the best estimate of the *median* percentage of households that had computers during the years shown?

   A 30% C 50%
   B 63% D 58%

4. Use the trends shown in this graph to predict the percent of households that will have computers in 2004.

   F 42% H 50%
   G 60% J 72%

Use the circle graph for Numbers 5–8.

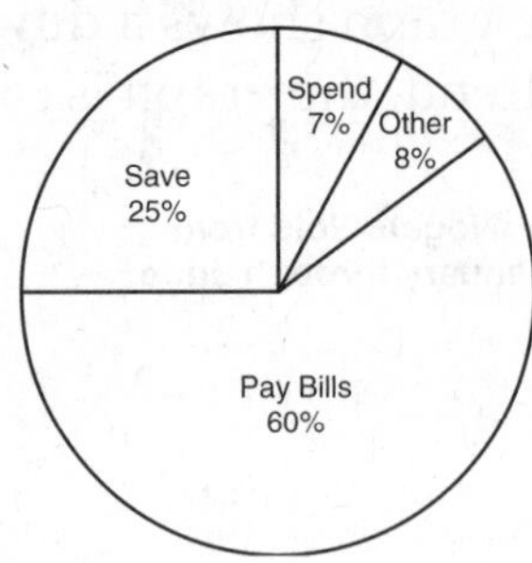

5. Which question would get the most unbiased response for this survey?

   A What do you plan to do with your tax refund?
   B What do you plan to buy with your tax refund?
   C How much of your tax refund will you use to pay bills?
   D How much of your tax refund do you plan to put into savings?

6. 1,000 people were surveyed. How many said they would spend their refund?

   F 300 H 100
   G 70 J 250

7. Approximately what fraction of the people surveyed would save the money from their tax refund?

   A $\frac{2}{3}$ C $\frac{1}{2}$
   B $\frac{1}{4}$ D $\frac{3}{5}$

8. Based on the percentages shown, if 5,000 people were surveyed, and you selected one of them at random, what is the probability that the person said he or she would pay bills with the refund?

   F $\frac{3}{5}$ H $\frac{2}{3}$
   G $\frac{3}{4}$ J $\frac{5}{6}$

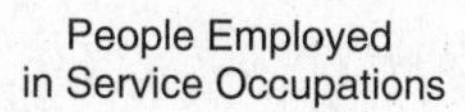

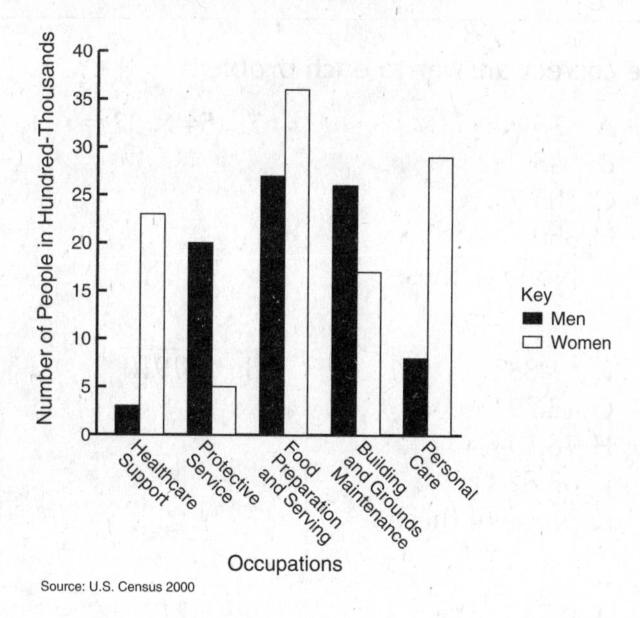

Use the above graph to answer Numbers 9–14.

9. Which category shows about 4 times as many men employed as women?
   - **A** healthcare support
   - **B** protective service
   - **C** food preparation and serving
   - **D** building and grounds maintenance

10. In which category is the number of women employed about $\frac{2}{3}$ the number of men?
    - **F** healthcare support
    - **G** protective service
    - **H** food preparation and serving
    - **J** building and grounds maintenance

11. In which category are the least number of men employed?
    - **A** healthcare support
    - **B** protective service
    - **C** food preparation and serving
    - **D** building and grounds maintenance

12. Approximately how many more women than men are employed in preparing and serving food?
    - **F** 1 million
    - **G** 100 thousand
    - **H** 10 thousand
    - **J** 1 thousand

13. In which category are the greatest number people employed?
    - **A** health support
    - **B** personal care
    - **C** food preparation and serving
    - **D** protective service

14. Which is the best estimate of the total number of people employed in service occupations?
    - **F** 19 hundred
    - **G** 19 thousand
    - **H** 19 hundred thousand
    - **J** 19 million

Skills Inventory Posttest

Part A: Computation

Circle the letter for the correct answer to each problem.

1. $245 \div 7 =$
   A 35
   B 45
   C 305
   D 350
   E None of these

2. $12.12 + 6.7 =$
   F 1.882
   G 12.79
   H 18.19
   J 18.82
   K None of these

3. $324 \times 21 =$
   A 9,720
   B 6,804
   C 972
   D 6,704
   E None of these

4. $4\overline{)856}$
   F 216
   G 214
   H 224
   J 211 R2
   K None of these

5. $\begin{array}{r} 230 \\ \times\ 50 \\ \hline \end{array}$
   A 11,500
   B 1,150
   C 10,500
   D 1,050
   E None of these

6. $255 \times 6 =$
   F 1,500
   G 1,230
   H 1,530
   J 3,680
   K None of these

7. $54 \times 32 =$
   A 1,638
   B 1,628
   C 270
   D 1,728
   E None of these

8. $13\overline{)910}$
   F 7
   G 70
   H 6
   J 77
   K None of these

9. $6{,}212 \div 2 =$
   A 316
   B 3,016
   C 3,101
   D 311
   E None of these

10. $6\frac{5}{8} - 2\frac{3}{8} =$
    F $\frac{1}{4}$
    G $4\frac{1}{8}$
    H $4\frac{1}{4}$
    J $4\frac{1}{2}$
    K None of these

11. $12\overline{)16.8}$
    A 14
    B 1.4
    C 0.14
    D 10.4
    E None of these

12. $32 - 0.017 =$
    F 31.083
    G 31.83
    H 14
    J 30.3
    K None of these

13. $12.00 − 0.17

A $11.83
B $11.93
C $12.93
D $12.17
E None of these

14. $\frac{11}{15} + \frac{4}{15}$

F $\frac{7}{15}$
G $\frac{1}{2}$
H 1
J 15
K None of these

15. $\frac{3}{4} \div 3 =$

A $\frac{1}{4}$
B $2\frac{1}{4}$
C 4
D $\frac{3}{4}$
E None of these

16. 45% of 200 =

F 90
G 900
H 18
J 180
K None of these

17. $0.37 \times 3 =$

A 11.1
B 1.11
C 111
D 0.111
E None of these

18. $17 - (-5) =$

F 12
G 22
H −12
J −22
K None of these

19. 15% of ☐ = 18

A 27
B 120
C 180
D 150
E None of these

20. $\frac{-36}{-90} =$

F $\frac{2}{5}$
G $-\frac{2}{5}$
H $\frac{4}{9}$
J $-\frac{4}{5}$
K None of these

21. What percent of 150 is 75?

A 75%
B 50%
C 20%
D 12%
E None of these

22. $-5 + -3 + 2 =$

F −10
G 0
H −6
J 10
K None of these

23. $\frac{3}{5} \times \frac{2}{3} =$

A $\frac{1}{10}$
B $\frac{9}{10}$
C $2\frac{1}{5}$
D $\frac{2}{5}$
E None of these

24. 20% of ☐ = 60

F 30
G 120
H 80
J 12
K None of these

25. $5 - (-1) =$

A 4
B −4
C 6
D −6
E None of these

Part B: Applied Mathematics

Circle the letter for the correct answer to each problem.

1. The annual city budget of Centerville is \$2.5 million. Which of these is the same as 2.5 million?

   A $2\frac{1}{2}$ million **C** 2,005,000

   B $2\frac{1}{4}$ million **D** 205 thousand

2. In which of these equations is x equal to 9?

   F $x + 9 = 9$
   G $9 - x = 0$
   H $9 \times x = 18$
   J $4 + x = 9$

3. Which of these fractions, if any, is greater than 1?

   A $\frac{3}{4}$ **C** $\frac{12}{15}$

   B $\frac{3}{2}$ **D** None of these

4. Which of these numbers is thirty thousand, four hundred twelve?

   F 3,412
   G 34,012
   H 30,412
   J 3,402

5. A box containing 15 marigold plants costs \$3. How much would 25 plants cost?

   A \$4
   B \$4.50
   C \$4.75
   D \$5

6. One glass goblet costs \$14.95. Which of these is the best estimate of how much 21 goblets would cost?

   F \$300
   G \$210
   H \$280
   J \$200

7. 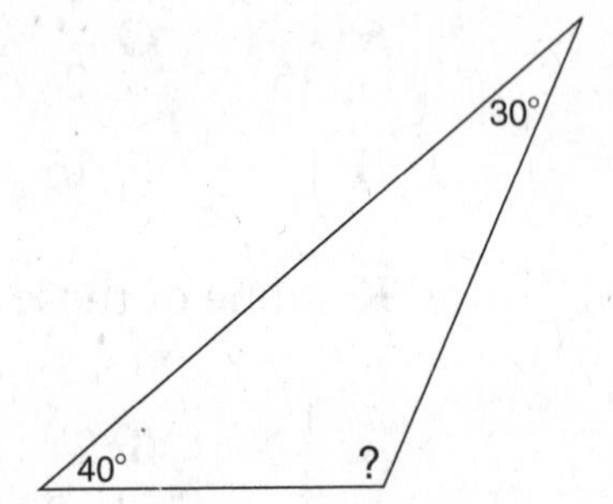

   What is the measure of the unlabeled angle in the triangle above?

   A 90°
   B 100°
   C 110°
   D There is no way to tell.

8. Frank's cable bill is \$27.50 per month. Which of these number sentences could be used to find how much he pays for cable in one year?

   F \$27.50 + 12 = □
   G \$27.50 ÷ 12 = □
   H \$27.50 × 12 = □
   J \$27.50 + \$27.50 = □

This list describes the installment plan at a furniture store. Study the description. Then do Numbers 9–12.

> Moore's Furniture Installment Plan
> - 15% down
> - 18% annual interest
> - No payments for 6 months
>
> Minimum monthly payment:
>
> Only $\frac{1}{10}$ of your balance

9. Sean decides to buy a bedroom set for $3,190, including tax. If he puts 15% down, how much will he have left to pay?

   A $478.50   C $2,711.50
   B $47,850   D $2,712.50

10. Sean's bedroom set was priced at $2,900, and he paid $290 in tax. What tax rate was he charged?

    F 29%
    G 2.9%
    H 10%
    J 19%

11. Which of these fractions shows how much interest the store charges each year?

    A $\frac{18}{1}$ C $\frac{18}{10}$
    B $\frac{9}{50}$ D $\frac{1}{18}$

12. Which of these shows how long Sean can wait before making his first payment?

    F $\frac{1}{6}$ of a year   H $1\frac{1}{2}$ years
    G $\frac{2}{3}$ of a year   J $\frac{1}{2}$ of a year

This list shows the ingredients needed to make 4 servings of pancakes. Study the list. Then do Numbers 13–16.

Pancakes

$\frac{1}{2}$ cup flour	$\frac{1}{4}$ tsp salt
$\frac{1}{2}$ tsp sugar	1 egg
$\frac{1}{2}$ tsp baking powder	
$\frac{1}{4}$ to $\frac{1}{2}$ cup milk	

13. This recipe calls for twice as much sugar as which ingredient?

    A baking powder
    B salt
    C milk
    D flour

14. Which of these could be a proper quantity of milk for 4 servings of pancakes?

    F $\frac{1}{5}$ cup   H $\frac{2}{3}$ cup
    G $\frac{3}{4}$ cup   J $\frac{1}{3}$ cup

15. Dan is going to triple the recipe above. How many servings of pancakes will he make?

    A 7   C 12
    B 14   D 15

16. How many cups of flour will Dan need to triple the recipe?

    F $\frac{1}{6}$ H $1\frac{1}{2}$
    G 2   J $\frac{1}{5}$

Mrs. Mahon ordered a hat rack from a catalog. The parts need to be assembled, and the diagram below shows all the pieces that came with her order. Study the diagram. Then do Numbers 17–21.

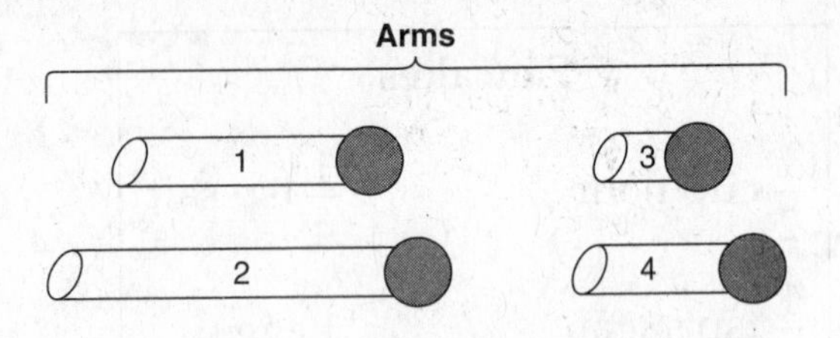

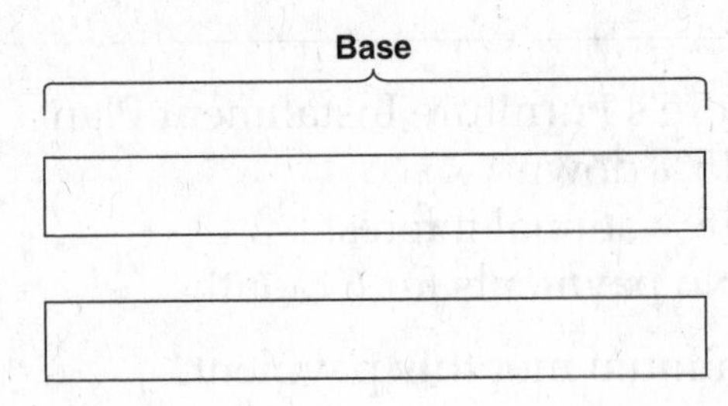

Stand

4 1/2 feet

17. Which two parts of this diagram are congruent?

    A Arm 3 and Arm 4
    B Arm 1 and Arm 2
    C the pieces of the base
    D the arms and the stand

18. The directions say that Arm 1 should be $\frac{1}{4}$ of the way down from the top of the stand. How many feet from the top of the stand should Arm 1 be?

    F 2 feet
    H $1\frac{1}{8}$ feet
    G $1\frac{1}{2}$ feet
    J $1\frac{1}{4}$ feet

19. What shape is formed by the rounded end of each arm?

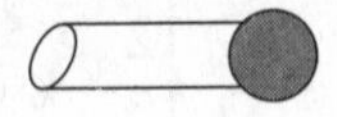

    A a sphere
    B a parallelogram
    C a cone
    D a circle

20. The directions say that there should be a 30° angle between each arm and the stand. Which of these shows an angle that is about 30°?

F
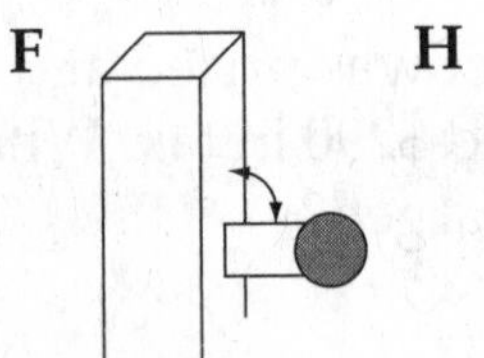

H
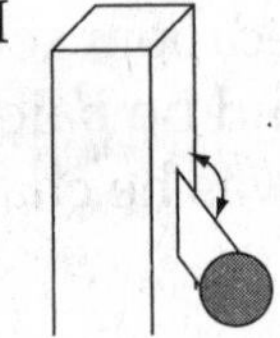

G
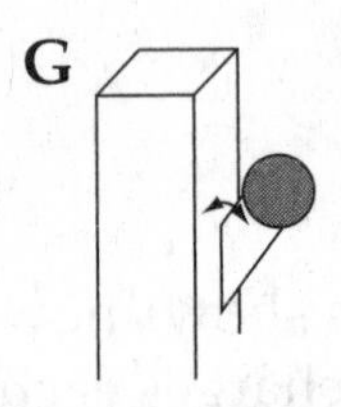

J None of these

21. The directions say that the two pieces of the base should be nailed together so that one piece is perpendicular to the other. Which of these shows how the base should look?

A C
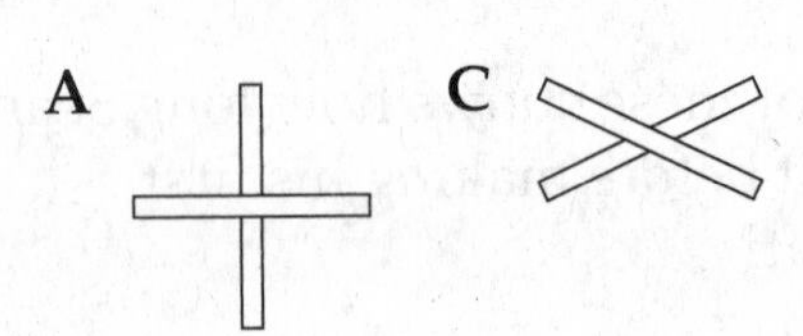

B
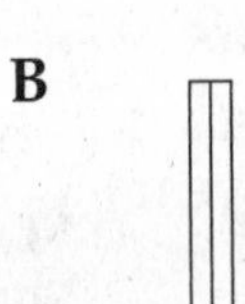

D
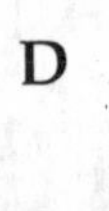

This graph shows the average number of years different types of pets live. It also shows the longest possible life each type of pet could have. Study the graph. Then do Numbers 22–26.

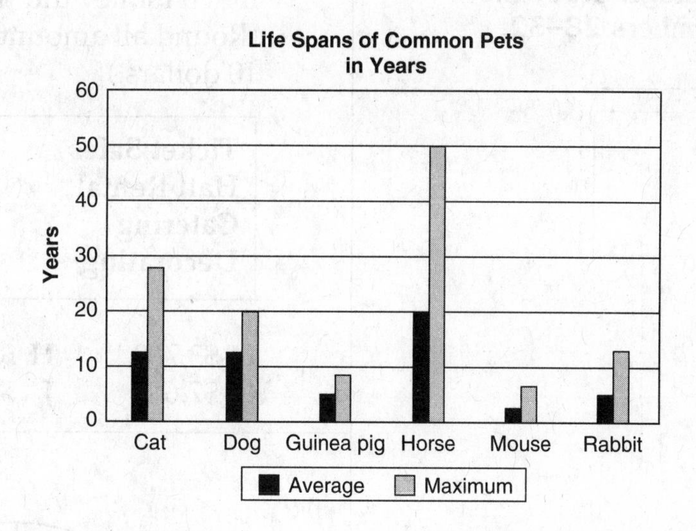

22. Which of the animals listed has the longest average lifespan?

    F cat
    G guinea pig
    H horse
    J rabbit

23. The oldest cats live 28 years. Approximately how many times longer is that than the average cat's lifespan?

    A half as long
    B twice as long
    C 4 times as long
    D 3 times as long

24. Which of the following statements is supported by the information in this graph?

    F Animals live longer in the wild than they do in captivity.
    G Pets tend to live longer than livestock.
    H Many pets die before their time.
    J The smaller an animal is, the shorter its life span tends to be.

25. Which two pets, if any, have about the same average life span?

    A cats and dogs
    B mice and rabbits
    C horses and dogs
    D none of them

26. Which animal has an average life span that is about half as long as that of the average horse?

    F cat or dog
    G mouse
    H guinea pig
    J rabbit

27. Karin asks Jo to pick a number between 0 and 10. What is the probability that the number Jo picks will be even?

    A $\frac{1}{2}$ C $\frac{3}{4}$

    B $\frac{4}{9}$ D $\frac{2}{3}$

The Brookfield Women's Shelter is holding a fashion show to raise money. This diagram shows their stage. Study the diagram. Then do Numbers 28–32.

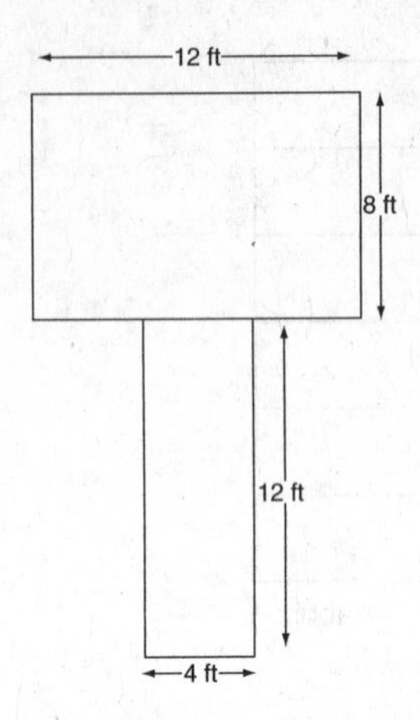

28. How much carpeting would you need to cover the entire stage?

    F 72 square feet
    G 96 square feet
    H 144 square feet
    J 48 square feet

29. The steps leading up to the stage will each be 9 inches tall. The stage is 3 feet tall. How many steps will there be leading up to the stage?

    A 3 C 4
    B 5 D 6

30. There will be cloth tacked around the perimeter of the stage to hide all the supports underneath. How much cloth will be needed?

    F 36 feet H 64 feet
    G 68 feet J 56 feet

31. There will be 28 outfits modeled during the show, and each outfit will be shown for about 45 seconds. About how long will the show last?

    A 126 minutes C 2 hours
    B 21 minutes D 45 minutes

32. This balance sheet lists all the show's income and expenses. About how much money did the show make? (Round all amounts to the nearest 10 dollars.)

Ticket Sales	**+$5,215.00**
Hall Rental	**−$450.00**
Catering	**−$1,625.00**
Decorating	**−$375.95**

    F $3,760 H $2,760
    G $7,680 J $2,770

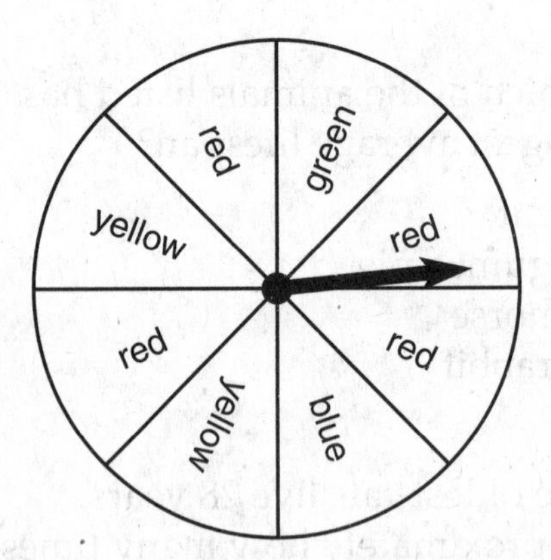

Use the spinner for Numbers 33 and 34.

33. Marion spins the spinner. What is the probability that it will land on red?

    A $\frac{3}{8}$ C $\frac{1}{4}$
    B $\frac{3}{4}$ D $\frac{1}{2}$

34. Which two colors have the same probability of having the spinner land on them?

    F blue and yellow
    G red and yellow
    H green and blue
    J yellow and green

Diane and Jarrad made resolutions to lose weight. This graph shows how well they did on their diets during the first month. Study the graph. Then do Numbers 35–38.

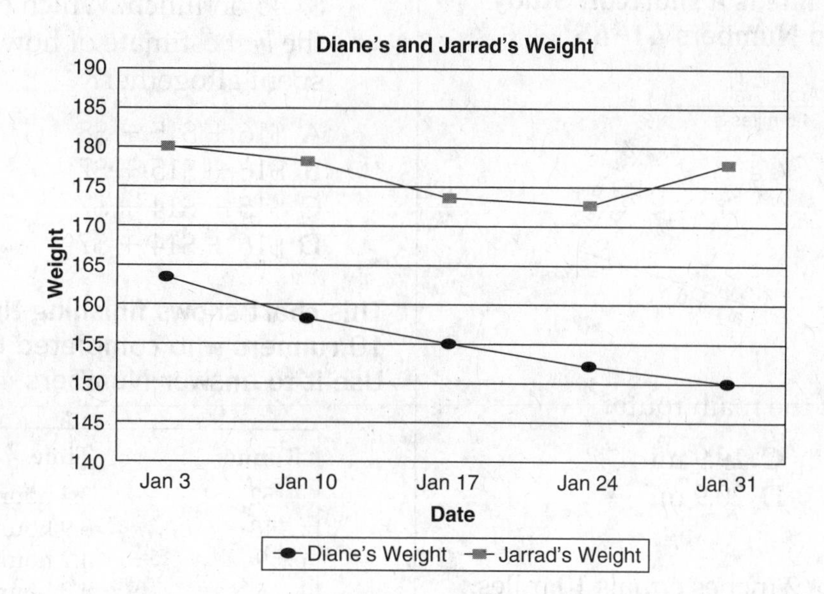

35. In which week did Jarrad lose the greatest amount of weight?

 A January 3 through January 10
 B January 10 through January 17
 C January 17 through January 24
 D January 24 through January 31

36. On which of the dates shown was there the greatest difference between Diane's and Jarrad's weights?

 F January 3
 G January 10
 H January 24
 J January 31

37. On average, about how much did Diane lose each week in January?

 A $3\frac{1}{2}$ pounds **C** 5 pounds
 B 6 pounds **D** 8 pounds

38. If she continues to lose weight at this rate, how much will Diane weigh at the end of March?

 F 163 pounds
 G 135 pounds
 H 122 pounds
 J 100 pounds

39. Rena gets 20% off at a clothing store where she works. She bought 3 scarves that were marked $24.95 each. About how much did she save in all? (Round to the nearest dollar.)

 A $5
 B $6
 C $15
 D $18

40. Ron is a fireman. He always works 3 days in a row, then takes off 4 days in a row. This month, he worked on the 3rd, 4th, and 5th. Which of the following days this month will he be off?

 F 10 **H** 14
 G 12 **J** 17

Every year the city of Munro hosts a race. This map shows the route that runners take. The dotted line is a shortcut. Study the map. Then do Numbers 41–45.

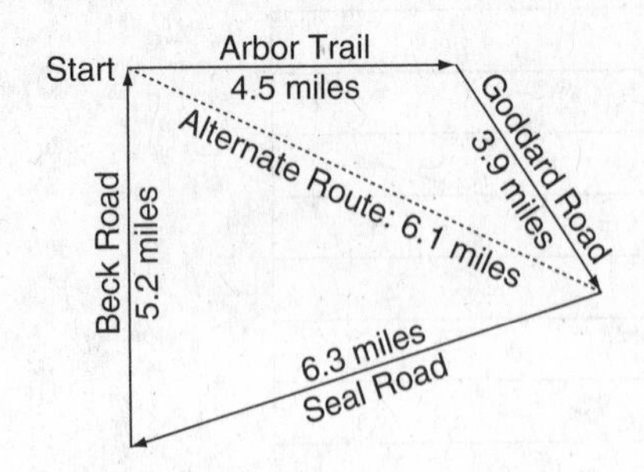

41. How long is the main route?

    A 18.9 mi C 19.9 mi
    B 20.1 mi D 20.9 mi

42. On this map, 2 inches equals 10 miles. What scale was used to draw the map?

    F 1 inch = 1 mile
    G 1 inch = 2 miles
    H 1 inch = 5 miles
    J 1 inch = 20 miles

43. Sheila took Arbor Trail and Goddard Road but then took the alternate route to the finish line. In which of these equations is x the number of miles by which she shortened the main route?

    A $4.5 + 3.9 + 6.1 = x$
    B $6.3 + 5.2 = x$
    C $(6.3 + 5.2) - 6.1 = x$
    D $\frac{6.3 + 5.2}{6.1} = x$

44. It took Sheila 47 minutes to run the length of the Arbor Trail. About how fast was she running?

    F 5 miles an hour
    G 10 miles an hour
    H 6 miles an hour
    J 4.5 miles an hour

45. At the race, Sheila spent $15.95 on her entrance fee, $14.72 on a T-shirt, and $7.45 on lunch. Which expression gives the *best* estimate of how much she spent altogether?

    A $16 + $15 + $8
    B $16 + $15 + $7
    C $15 + $14 + $7
    D $16 + $14 + $7

This chart shows finishing times of the first 10 runners who completed the full route. Use it to answer Numbers 46 and 47.

Runner	Time
52	3.4 hours
13	3.9 hours
9	4.2 hours
5	4.3 hours
11	4.6 hours
18	4.7 hours
27	4.7 hours
32	4.8 hours
51	5.1 hours
15	5.3 hours

46. Which of the following is the runners' mean or average finishing time?

    F 3.5 hours
    G 4.5 hours
    H 4.1 hours
    J 4.8 hours

47. If one of the runners listed was selected at random, what is the probability that his or her time was less than $4\frac{1}{2}$ hours?

    A $\frac{2}{5}$ C $\frac{2}{3}$
    B $\frac{1}{2}$ D $\frac{1}{4}$

48. Round each of the following numbers to the nearest whole number.

1.12	0.89	$1\frac{3}{4}$	$\frac{1}{3}$

How many of the numbers round to 1?

F 1
H 2
G 3
J 4

49. The chart below shows how a function changes each "Input" number to a corresponding "Output" number. What number is missing from the table?

Input	5	1	$2\frac{1}{2}$	$5\frac{1}{2}$
Output	$6\frac{1}{2}$	$2\frac{1}{2}$	4	

A 6
C $6\frac{1}{2}$
B $3\frac{1}{2}$
D 7

50. Karin has to print 45 copies of a 50-page report. Paper is sold in 500-sheet packages. To find out how many packages of paper she needs, Karin uses this equation:

$$45 \times 50 = n$$

What does n represent in this equation?

F how many sheets of paper she needs
G how many packages of paper she needs
H how many reports she can print with one package of paper
J how many sheets of paper are in 4 packages of paper

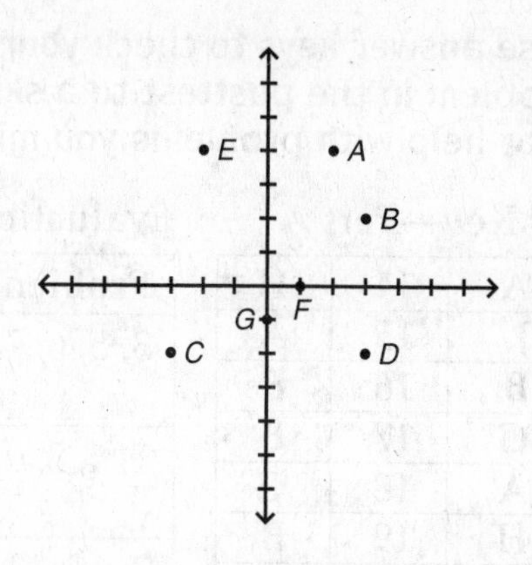

Use the coordinate grid shown to answer Numbers 51 and 52.

51. Which pair of coordinates represents the location of Point *A*?

A (3, −2)
B (−2, 3)
C (4, 2)
D (2, 4)

52. What point is found at (1, 0)?

F point *G*
G point *F*
H point *E*
J point *D*

Posttest Answer Key and Evaluation Chart

Use these answer keys to check your answers on the Posttest. The evaluation chart matches each problem in the posttest to a skill area. Use the charts to find which pages in this book to use for help with problems you missed.

Answer Key—Part A

1	A	14	H
2	J	15	A
3	B	16	F
4	G	17	B
5	A	18	G
6	H	19	B
7	D	20	F
8	G	21	B
9	E	22	H
10	H	23	D
11	B	24	K
12	K	25	C
13	A		

Evaluation Chart—Part A

Problem Number	Skill Area	Text Pages
3, 5, 6, 7	Multiplication of Whole Numbers	18–22
1, 4, 8, 9	Division of Whole Numbers	23–29
2, 11, 12, 13, 17	Decimals	40–50, 57
10, 14, 15, 23	Fractions	53–70
18, 20, 22, 25	Integers	73–78
16, 19, 21, 24	Percents	88–96

Answer Key—Part B

1	A	27	B
2	G	28	H
3	B	29	C
4	H	30	H
5	D	31	B
6	F	32	H
7	C	33	D
8	H	34	H
9	C	35	B
10	H	36	J
11	B	37	A
12	J	38	H
13	B	39	C
14	J	40	H
15	C	41	C
16	H	42	H
17	C	43	C
18	H	44	H
19	A	45	B
20	G	46	G
21	A	47	A
22	H	48	H
23	B	49	D
24	J	50	F
25	A	51	D
26	F	52	G

Evaluation Chart—Part B

Problem Number	Skill Area	Text Pages
1, 2, 3, 4, 5, 8, 10, 11, 14, 42	Number and Number Operations	8–16, 18–21, 23–26, 40–41, 53–59, 81–95
13, 15, 16, 41	Computation in Context	43–50, 60–69
6, 32, 44, 45, 48	Estimation	17, 22, 27, 37, 42
12, 28, 29, 30, 51, 52	Measurement	129–148
22, 23, 24, 25, 26, 35, 36	Data Analysis	154–157, 159–163
27, 33, 34, 37, 46, 47	Statistics and Probability	151–153, 158
40, 43, 49, 50	Patterns, Functions, Algebra	102–109
7, 17, 19, 20, 21	Geometry and Spatial Sense	113–126
9, 18, 31, 38, 39	Problem Solving	32–36, 50, 70, 96, 110

Answer Key

The Number System-Whole Numbers

Page 8

1. ten thousands, thousands, hundreds, tens, ones
2. 1 hundred thousands, 7 thousands, 4 hundreds, 3 ones
3. 1 million, 5 hundred thousands
4. ten thousands
5. hundred thousands
6. 2
7. 5
8. 5 million
9. 6 hundred thousand
10. 7 thousand
11. 3 hundred thousand
12. 50 thousand

Page 9

1. fifty-nine thousand, five hundred five
2. two hundred eleven thousand, five hundred twenty-five
3. four million, five hundred thousand, seventy-two
4. nine hundred one thousand, three hundred sixteen
5. twenty-three thousand, twelve
6. 91,201
7. 63,412
8. 3,400,000
9. 801,356

Page 10

1. 12,301
2. 946,203
3. 55,000
4. 112
5. 91,415
6. 1,000,000
7. 12,000
8. 45,678
9. 1,512
10. 999,999
11. 1,042
12. 899
13. 812
14. 750
15. 1,596
16. 654
17. 52, 115, 269, 397
18. 952; 1,736; 6,710; 9,000
19. 952; 978; 5,107; 5,123; 17,008
20. 8,056; 1,000; 425; 99
21. 4,000; 3,985; 3,280; 997
22. 1,840; 1,480; 1,408; 1,048
23. 9,210
24. 8,651
25. 7,321
26. 2,035
27. 1,678
28. 2,034

Page 11

Numbers 2, 3, and 6 call for an estimate.

7. B
8. H
9. A
10. H
11. B
12. H

Page 12

1. 180
2. 810
3. 100
4. 0
5. 400
6. 1,200
7. 0
8. 1,000
9. $46.00
10. $20.00
11. $1.00
12. $128.00
13. 3,100
14. 12,000
15. $3.00
16. $12.60
17. $90
18. 1,000

The Number System Skill Checkup

Pages 13–14

1. B
2. H
3. D
4. J
5. B
6. H
7. C
8. J
9. B
10. H
11. B
12. H
13. D
14. F
15. B
16. J
17. D
18. H
19. C

Computation with Whole Numbers

Page 15

1. 1,061
2. 84,767
3. 673 gal
4. 566
5. 12,901
6. $18.73
7. 120 min
8. 5,402
9. 1,426
10. 17,060

11. 214,331
12. 1,517
13. 1,747
14. 1,273
15. 421
16. 363

Page 16

1. 223 miles
2. 1,250
3. 3,120
4. $37.50
5. 537
6. 2,250
7. 116
8. $12.24
9. 133
10. 1,824
11. 11,498
12. 2,858
13. 47,984
14. 26

Page 17

Answers may vary depending on how the numbers are rounded.

1. 900
2. 2,400
3. 21,000
4. 24,000
5. 2,300
6. 97,000
7. 20,000
8. 50
9. 240
10. 42,000
11. 5,500
12. 593,000
13. 1,600
14. 7,000

Page 18

1. 248
2. 930
3. 2,555
4. 486
5. 808
6. $8.42
7. $14.08
8. 24,408
9. $168
10. 1,569
11. 300
12. 639
13. 1,688
14. $3.96
15. 1,206
16. $15.50

Page 19

1. 3,020
2. 623
3. 207
4. 30,054
5. 175
6. 2,325
7. 5,260
8. 7,250
9. 4,059
10. 322,888
11. 2,868
12. 8,118
13. 24,765
14. 4,004
15. 504
16. 31,120
17. 36,189
18. 8,190

Page 20

1. 3,840
2. 960
3. 3,720
4. 7,490
5. 9,600
6. 73,500
7. 7,500
8. 73,200
9. 940
10. 81,000
11. 16,260
12. 730
13. 350
14. 860
15. 450
16. 28,080
17. 370
18. 840
19. 960
20. 4,140
21. 11,750
22. 3,480
23. 6,090
24. 13,800
25. 1,820
26. 27,150
27. 216,000
28. 105,000
29. 135,000

Page 21

1. 2,320; 2,726
2. 1,840; 1,978
3. 110, 1,100; 1,210
4. 11,808
5. 38,010
6. 30,340
7. 1,200
8. 3,690
9. 15,834
10. 9,592
11. 46,494
12. 15,400
13. 48,750

Page 22

Answers may vary.

1. 9,000
2. 8,000
3. 12,000
4. 1,200
5. 180,000
6. 6,000
7. 35,000
8. 4,000
9. 4,800
10. 6,000

11. 6,000
12. 500,000
13. 3,500
14. 8,000
15. 4,000
16. 18,000
17. 40,000
18. 2,000

Page 24

1. 5
2. 7
3. 6
4. 9
5. 7
6. 6
7. 9
8. 5
9. 7
10. 6
11. 4
12. 4
13. 5
14. 4
15. 5
16. 7
17. 7
18. 3
19. 2
20. 3
21. 6
22. 3
23. 8
24. 9
25. 8
26. 3
27. 4
28. 5
29. 8
30. 6

Page 25

1. 124
2. 31
3. 123
4. 102
5. 42
6. 108
7. 203
8. 40
9. 61
10. 104
11. 100
12. 109
13. 314
14. 21
15. 104
16. 12

Page 26

1. 13 R1
2. 10 R2
3. 11 R2
4. 2 R1
5. 10 R5
6. 47 R2
7. 102 R2
8. 103 R4
9. 109 R4
10. 203 R5
11. 52 R4
12. 147 R3
13. 140
14. 918
15. 216 R1
16. 840

Page 27

1. 100
2. 600
3. 200
4. 800
5. 60
6. 700
7. 100
8. 20
9. 4,000
10. 7,000
11. 700
12. 400

Page 29

1. 13
2. 24
3. 50
4. 12
5. 20
6. 16
7. 15 R3
8. 24 R4
9. 30
10. 22
11. 6
12. 30
13. 5
14. 4
15. 30 R20
16. 23
17. 50
18. 201
19. 35
20. 8
21. 402
22. 22
23. 402
24. 6
25. 30
26. 4
27. 32
28. 42 R1
29. 7 R1
30. 152

Computation with Whole Numbers Skills Checkup

Pages 30–31

1. D
2. F
3. B
4. J
5. E
6. F
7. B
8. F
9. C

10. F
11. E
12. J
13. B
14. H
15. A
16. B
17. H
18. B
19. H
20. D

Page 33

1. A
2. H
3. C
4. F
5. C

Problem Solving

Page 34

1. Becky agreed to make 12 dozen cookies. Her sister made 4 dozen.
2. 250 people have moved to Carville.
3. The hotel would cost $80 per person per night.
4. Ron is 65 years old. His wife is 3 years younger.
5. What is the rent per day?
6. How many lunches and how much did each lunch cost?
7. What time did he start and what time did he arrive, or, how many miles did he drive and what was his speed?
8. How many square feet does she need?

Page 35

2. H; 20 × 23¢ = $4.60
3. B; 89 − 36 = 53
4. G: $7,000 − $3,259 = $3,741
5. A; $48.58 + $3.60 = $52.18
6. J; $94 ÷ 4 = $23.50

Page 36

1. B
2. F
3. C
4. 45 min
5. $14.75
6. $3,026
7. 11

Page 37

1. C
2. H
3. B
4. H

Estimates may vary

5. 60
6. $320
7. $18
8. $6

Problem-Solving Skills Checkup

Pages 38–39

1. B
2. G
3. C
4. B
5. J
6. C
7. G
8. A
9. J
10. A
11. G
12. B
13. G
14. D

Decimals

Page 40

1. hundredths
2. tenths
3. hundredths
4. thousandths
5. 6.003
6. 0.100
7. 2.03
8. 0.012
9. two and four hundredths
10. one hundred forty thousandths
11. sixty-two and two hundredths
12. fifteen and two hundred seven thousandths
13. nine and nine hundredths

Page 41

1. 0.1
2. 0.13
3. 1.123
4. 12.50
5. 0.005
6. 1.013
7. 0.98
8. 4.014
9. 34.478; 34.487; 37.448; 38.744
10. 6.873; 9.909; 10.50; 12
11. 8.03; 8.035; 8.3; 8.35
12. 0.79; 0.865; 0.897; 0.9
13. 4.20; 4.051; 4.020; 4.015
14. 0.9; 0.63; 0.623; 0.20

Page 42

1. 0.8
2. 3.9
3. 7.0
4. 1.0
5. 16.0
6. 10.0
7. 0.37
8. 3.00
9. 14.00
10. 12.48
11. 6.78
12. 10.10
13. 2.005
14. 7.496
15. 0.012
16. 15.828
17. 0.990

18. 1.000
19. 4.38
20. 1.00
21. 0.99

Page 43

1. 32.01
2. 0.23
3. 1.523
4. 1.59
5. 0.105
6. 1.425
7. 7.40
8. 9.23
9. 23.209
10. 12.231

Page 44

1. 3
2. 0.762
3. 1.485
4. 0.43
5. 3.15
6. 15.575
7. $31.12
8. 68 cents
9. 0.234
10. 8.2
11. $15.58
12. 4.983

Page 45

1. 0.28
2. 0.21
3. 7.5
4. .603
5. 8.48
6. 0.1104
7. 1.2
8. 3.12
9. 0.2625
10. 76.26
11. 1.12
12. 14.4
13. 2.3
14. 1,352
15. 0.4
16. 14
17. 1.25
18. 90

Page 46

19. 0.08
20. 0.012
21. 0.038
22. 0.00125
23. 0.00186
24. 0.00048
25. 0.0035
26. 0.36
27. 0.2412
28. 0.006
29. 0.05
30. 0.027
31. 10
32. 0.006

Page 47

1. 2.04
2. 0.063
3. 16.1
4. $2.04
5. 30.09
6. 0.105
7. 0.5
8. 1.023
9. 5.3
10. 0.304
11. 0.126
12. 0.27
13. 0.0081
14. $7.14
15. 23

Page 48

1. 96
2. 6
3. 9
4. 49.6
5. 4.8
6. 24
7. 2.3
8. 26.3
9. 770
10. 714
11. 5.5
12. 1.22

Page 49

13. 400
14. 325
15. 60
16. 12,000
17. 6,300
18. 2,700
19. 320
20. 2,870
21. 120
22. 300

Page 50

1. D; $6.64
2. H; 114.4
3. B; 1.6 lbs
4. F; $15.42
5. 558.2 mi
6. 3.9 mi
7. $5.84
8. 2 cents
9. 24 min
10. $42.70

Decimals Skills Checkup

Pages 51–52

1. B
2. H
3. D
4. F
5. B
6. K
7. B
8. G
9. D
10. F
11. B
12. F
13. A

14. J
15. B
16. J
17. D
18. G
19. B
20. J

Fractions

Page 53

1. $\frac{17}{100}$
2. $2\frac{1}{2}$
3. $\frac{23}{59}$
4. $\frac{5}{12}$
5. $\frac{8}{12}$ or $\frac{2}{3}$
6. $\frac{18}{16}$ or $1\frac{1}{8}$

Page 54

1. >
2. <
3. >
4. <
5. <
6. $\frac{1}{9}, \frac{1}{7}, \frac{1}{5}$
7. $\frac{2}{9}, \frac{2}{5}, \frac{2}{3}$
8. $\frac{4}{9}, \frac{4}{8}, \frac{4}{7}$
9. $\frac{1}{7}, \frac{3}{7}, \frac{5}{7}$
10. $\frac{2}{5}, \frac{3}{5}, \frac{4}{5}$
11. $\frac{2}{9}, \frac{4}{9}, \frac{7}{9}$

Page 55

1. 4
2. 4; 20
3. $\frac{6}{6}$
4. 5; $\frac{5}{15}$
5. 2
6. 4; 5
7. $\frac{3}{3}$
8. $\frac{6}{6}$; 1
9. $\frac{7}{7}$; 2
10. $\frac{3}{3}$; 4

Page 56

1. W
2. M
3. M
4. M
5. W
6. M
7. W
8. W
9. M
10. W
11. $1\frac{1}{3}$
12. 2
13. $1\frac{1}{4}$
14. 5
15. $2\frac{3}{8}$
16. $1\frac{1}{2}$
17. $1\frac{2}{5}$
18. 2
19. $1\frac{4}{5}$
20. $3\frac{3}{5}$

Page 57

1. $\frac{4}{10} = \frac{2}{5}$
2. $\frac{45}{100} = \frac{9}{20}$
3. $\frac{8}{100} = \frac{2}{25}$
4. $\frac{125}{1{,}000} = \frac{1}{8}$
5. $\frac{60}{100} = \frac{3}{5}$
6. $3\frac{5}{10} = 3\frac{1}{2}$
7. $\frac{5}{100} = \frac{1}{20}$
8. $\frac{75}{100} = \frac{3}{4}$
9. 0.2
10. 0.375
11. 0.03
12. 0.32
13. 0.9
14. 0.35
15. 0.75

Page 58

1. C, D
2. F, G, J
3. A, D
4. G, H, J

Page 59

1. 1, 2, 3, 6, 9, 18
2. 1, 3, 5, 15
3. 1, 2, 4, 5, 10, 20
4. 1, 2, 3, 4, 6, 8, 12, 24
5. 1, 2, 4, 7, 14, 28
6. 1, 3, 9
7. 1, 2, 4, 8
8. 1, 7
9. 9; $\frac{1}{2}$
10. 5; $\frac{3}{4}$
11. 6; $\frac{3}{4}$
12. 4; $\frac{2}{7}$

Page 60

1. $\frac{7}{9}$
2. 1
3. $1\frac{1}{3}$

4. $\frac{2}{3}$
5. $1\frac{1}{12}$
6. $1\frac{1}{2}$
7. $5\frac{2}{3}$
8. 6
9. $4\frac{5}{16}$
10. $\frac{1}{3}$
11. $\frac{1}{12}$
12. $\frac{2}{3}$
13. $\frac{1}{2}$
14. $1\frac{1}{5}$
15. 2
16. $1\frac{2}{7}$
17. $4\frac{1}{4}$
18. $4\frac{2}{5}$
19. 5

Page 61

1. 6
2. 8
3. 5
4. 7
5. 11
6. 5
7. 8
8. $\frac{7}{4}$
9. $\frac{1}{5}$
10. $1\frac{1}{4}$
11. $2\frac{5}{7}$
12. $1\frac{7}{8}$
13. $\frac{2}{3}$
14. $2\frac{1}{3}$
15. $8\frac{3}{5}$
16. $\frac{13}{15}$

Page 62

1. 5: 5, 10, 15, 20,…
   3: 3, 6, 9, 12, 15,…
   LCM 15
2. 4: 4, 8, 12, 16, 20, 24, 28, …
   7: 7, 14, 21, 28,…
   LCM 28
3. 9: 9, 18, …
   6: 6, 12, 18, …
   LCM 18
4. 8: 8, 16, 24, 32, 40, 48, …
   12: 12, 24, 36, 48, …
   LCM 24

Page 63

1. $<$
2. $<$
3. $>$
4. $>$
5. $=$
6. $<$
7. $\frac{7}{12}, \frac{2}{3}, \frac{3}{4}$
8. $\frac{1}{3}, \frac{7}{18}, \frac{5}{12}$
9. $\frac{1}{2}, \frac{3}{4}, \frac{7}{8}$
10. $\frac{1}{4}, \frac{3}{10}, \frac{2}{5}$
11. $\frac{5}{8}, \frac{7}{9}, \frac{5}{6}$
12. $\frac{9}{20}, \frac{3}{5}, \frac{7}{10}$

Page 64

1. $1\frac{7}{10}$
2. $1\frac{13}{28}$
3. $1\frac{1}{6}$
4. $\frac{2}{5}$
5. $\frac{17}{28}$
6. $\frac{2}{9}$
7. $\frac{1}{2}$
8. $\frac{1}{12}$
9. $1\frac{2}{9}$
10. $\frac{11}{45}$
11. $\frac{17}{35}$
12. $\frac{3}{8}$
13. $\frac{13}{24}$
14. 0
15. $1\frac{13}{24}$
16. $\frac{1}{4}$
17. $1\frac{1}{6}$

Page 65

1. $\frac{3}{20}$
2. $\frac{2}{15}$
3. $\frac{4}{9}$
4. $\frac{3}{25}$
5. 3
6. $\frac{2}{5}$
7. $\frac{3}{8}$
8. $\frac{4}{45}$
9. $1\frac{2}{5}$
10. $\frac{1}{6}$

11. $\frac{3}{20}$
12. $\frac{9}{35}$
13. $1\frac{1}{9}$
14. $1\frac{5}{7}$
15. 8
16. 10
17. $1\frac{1}{5}$
18. 15
19. $\frac{1}{8}$ cup
20. 10

Page 66

1. $\frac{2}{15}$
2. $\frac{1}{9}$
3. $\frac{1}{6}$
4. $\frac{1}{30}$
5. 8
6. $\frac{2}{5}$
7. $\frac{3}{5}$
8. $\frac{1}{5}$
9. $\frac{1}{7}$
10. $\frac{3}{7}$
11. 11
12. $\frac{1}{2}$ lb
13. $\frac{2}{5}$
14. 16
15. $\frac{5}{8}$

Page 67

1. $\frac{5}{4}$
2. $\frac{1}{7}$
3. $\frac{3}{4}$
4. $\frac{2}{7}$
5. $\frac{12}{5}$
6. $\frac{9}{1}$
7. $\frac{1}{6}$
8. $\frac{7}{2}$
9. $\frac{2}{9}$
10. $\frac{9}{6}$
11. $\frac{8}{1}$
12. $\frac{1}{15}$
13. $\frac{1}{2}$
14. $\frac{8}{9}$
15. $\frac{10}{9}$

Page 68

1. $\frac{2}{1}$; $1\frac{1}{2}$
2. $\frac{5}{2}$; $1\frac{1}{2}$
3. $\frac{3}{2}$; $\frac{2}{3}$
4. $\frac{4}{5}$; $\frac{8}{15}$
5. $2\frac{2}{5}$
6. $\frac{7}{16}$
7. 3
8. $2\frac{2}{15}$
9. $1\frac{5}{16}$
10. 1
11. $\frac{4}{9}$
12. 10
13. $1\frac{1}{6}$
14. $\frac{6}{7}$
15. $1\frac{1}{5}$
16. $1\frac{2}{5}$

Page 69

1. 18
2. 8
3. 12
4. $\frac{1}{7}$
5. $\frac{1}{6}$
6. $\frac{2}{45}$
7. $\frac{1}{15}$
8. $\frac{1}{21}$
9. $7\frac{1}{2}$
10. $2\frac{2}{3}$
11. $\frac{2}{3}$
12. 8
13. 10
14. 8

Page 70

1. B
2. H
3. D
4. G
5. 18
6. $80
7. $3\frac{17}{24}$ miles
8. $21
9. 21
10. Yes. $\frac{1}{3} > \frac{1}{4}$

Fractions Skills Checkup

Pages 71–72

1. B
2. J
3. C
4. K
5. C
6. K
7. C
8. H
9. B
10. H
11. D
12. F
13. D
14. H
15. C
16. A
17. H
18. C
19. G
20. C

Integers

Page 73

1. $<$
2. $<$
3. $>$
4. $>$
5. $<$
6. $<$
7. $>$
8. $=$
9. $^{-}12, 0, 3, 5$
10. $^{-}52, ^{-}13, 4, 12$
11. $^{-}1.5, -\frac{1}{2}, \frac{3}{4}, 3$

Page 74

1. 3.1
2. 3
3. 5
4. 5
5. 6
6. 0.25
7. 32
8. 0
9. $=$
10. $<$
11. $<$
12. $=$
13. $>$
14. $<$
15. $<$
16. $=$
17. $=$
18. $<$
19. $>$
20. $<$
21. $<$
22. $>$
23. $=$
24. $>$

Page 75

1. $^{-}3$
2. 4
3. $^{-}1$
4. 1
5. $^{-}5$
6. $^{-}2$
7. 0
8. $^{-}2$
9. 5
10. $^{-}15$
11. $^{-}4^{\circ}$ F
12. left

Page 76

1. 3
2. $^{-}17$
3. 19
4. 12
5. $^{-}11$
6. $^{-}3$
7. $^{-}7 + ^{-}4$
8. $14 + ^{-}6$
9. $^{-}8 + 4$
10. $3 + 9$
11. $5 + ^{-}12$
12. $13 + 2$
13. 5
14. $^{-}5$
15. 1
16. $^{-}3$
17. 6
18. 7
19. $^{-}5$
20. $^{-}10$
21. 9

Page 77

1. $^{-}6$
2. 20
3. $^{-}12$
4. $^{-}32$
5. $^{-}30$
6. 64
7. 35
8. $^{-}18$
9. $^{-}24$
10. $^{-}4$
11. $^{-}8$
12. 10
13. $^{-}2$
14. $^{-}2$
15. 4
16. 5
17. $^{-}7$
18. $^{-}16$
19. $\frac{1}{3}$
20. $-\frac{1}{2}$
21. $-\frac{2}{3}$

Page 78

1. $^{-}$\$72.41
2. 1ºC
3. 4 ft
4. 12,242 ft
5. \$168
6. \$24,080

7. 10 min
8. $2,325 profit
9. $^{-}8$
10. 5 floors

Integers Skills Checkup

Page 79 – 80

1. B
2. J
3. A
4. F
5. A
6. B
7. H
8. C
9. H
10. A
11. H
12. A
13. H
14. D
15. J
16. G
17. D
18. J
19. B
20. F

Ratio, Proportion, and Percent

Page 81

1. $\frac{\text{inches}}{\text{miles}} = \frac{1}{25}$
2. $\frac{\$}{\text{months}} = \frac{2{,}958}{1}$
3. $\frac{\text{evergreens}}{\text{leafy}} = \frac{1}{3}$
4. $\frac{\$}{\text{dozen}} = \frac{4}{1}$
5. $\frac{\text{hours}}{\text{placemats}} = \frac{1}{8}$
6. $\frac{\text{miles}}{\text{hours}} = \frac{450}{1}$
7. $\frac{\text{miles}}{\text{gallon}} = \frac{20}{1}$
8. $\frac{\text{men}}{\text{women}} = \frac{3}{2}$
9. $\frac{\text{games lost}}{\text{games played}} = \frac{3}{13}$
10. $\frac{\$\text{ materials}}{\$\text{ labor}} = \frac{17}{60}$

Page 82

1.

cents	3	6	9	12	15	18	21
cans	2	4	6	8	10	12	14

2.

length	5	10	15	20	25	30	35
width	2	4	6	8	10	12	14

Page 83

1. proportion, 16, 112
2. proportion, 8, 312, 312
3. not a proportion, 15, 5, 45, 35
4. not a proportion, 36, 64, 9, 144, 576
5. proportion, 27, 11, 9, 297, 297
6. proportion, 3, 72, 2, 108, 216, 216
7. N
8. N
9. Y
10. Y
11. N
12. Y

Page 84

1. centimeters; $\frac{\text{cm}}{\text{ft}}$; $\frac{1}{3} = \frac{n}{18}$
2. yards; $\frac{\text{yards}}{\$}$; $\frac{2}{8.95} = \frac{n}{53.70}$
3. days; $\frac{\text{feet}}{\text{days}}$; $\frac{15}{1} = \frac{60}{n}$
4. calories; $\frac{\text{cookies}}{\text{calories}}$; $\frac{2}{270} = \frac{5}{n}$
5. miles; $\frac{\text{miles}}{\text{minutes}}$; $\frac{1}{3} = \frac{n}{5}$
6. pages; $\frac{\text{minutes}}{\text{pages}}$; $\frac{15}{1} = \frac{n}{25}$
7. scripts turned down; $\frac{\text{scripts turned down}}{\text{scripts accepted}}$; $\frac{25}{1} = \frac{n}{8}$
8. yellow triangles; $\frac{\text{red triangles}}{\text{yellow triangles}}$; $\frac{3}{9} = \frac{21}{n}$

Page 85

1. example
2. 8, 14, 168, 14, 12
3. n, 15, 3, 120, 40
4. $n = 32$
5. $n = 2$
6. $n = 12$
7. $n = 36$
8. $n = 18$
9. $n = 12$
10. $n = 21$
11. $n = 18$
12. $n = 20$
13. $n = 150$
14. $n = 15$
15. $n = 12$

Page 86

1. 7 tiles
2. 18 packages
3. $40
4. $31
5. 120 miles
6. 5 cards

Page 87

1. B
2. C
3. F
4. J
5. B
6. C
7. G
8. H

Page 88

Percent	Fraction	Simplest Terms	Decimal
3%	$\frac{3}{100}$	$\frac{3}{100}$	0.03
5%	$\frac{5}{100}$	$\frac{1}{20}$	0.05
8%	$\frac{8}{100}$	$\frac{2}{25}$	0.08
10%	$\frac{10}{100}$	$\frac{1}{10}$	0.1
20%	$\frac{20}{100}$	$\frac{1}{5}$	0.2
50%	$\frac{50}{100}$	$\frac{1}{2}$	0.5
75%	$\frac{75}{100}$	$\frac{3}{4}$	0.75
80%	$\frac{80}{100}$	$\frac{4}{5}$	0.8

1. $\frac{3}{50}$
2. $\frac{3}{4}$
3. $\frac{2}{5}$
4. $\frac{3}{5}$

Page 89

1. 14%
2. 40%
3. 600%
4. 1.3%
5. 26%
6. 70%
7. 10%
8. 1%
9. 0.1%
10. 90%
11. 72%
12. 19%

Page 90

1. 10%
2. 60%
3. 50%
4. 15%
5. 25%
6. 12.5%
7. 37.5%
8. 20%
9. 80%
10. 62.5%
11. 50%
12. 20%

Page 91

1. 35
2. $20
3. 5
4. 30
5. 12.75 or $12\frac{3}{4}$
6. $48
7. $35
8. 9
9. 1
10. 7.8
11. 0.125
12. 0.005
13. 21 votes
14. $690

Page 92

1. $40
2. $25,200
3. $3.78
4. 6,775 votes

Page 93

1. A = 20%, B= 240, C = 48
2. D = 65%, E = 200, F = 130
3. A = 20%, B = 35, C = 7
4. D = 20%, E = 90, F = 18
5. A = 20%, B = $180, C = $36
6. D = 50%, E = 80, F = 40

Page 95

1. percent; $\frac{1{,}800}{9{,}000} = \frac{n}{100}$; $n = 20\%$
2. whole; $\frac{245}{n} = \frac{35}{100}$; $n = 700$ workers
3. percent; $\frac{5}{8} = \frac{n}{100}$; $n = 62.5\%$
4. part; $\frac{n}{1{,}200} = \frac{20}{100}$; $n = \$240$
5. percent; $\frac{6}{40} = \frac{n}{100}$; $n = 15\%$
6. whole; $\frac{5{,}240}{n} = \frac{25}{100}$; $n = 20{,}960$ people
7. part; $\frac{30}{100} = \frac{n}{2{,}460}$; $n = \$738$
8. percent; $\frac{48}{80} = \frac{n}{100}$; $n = 60\%$
9. whole; $\frac{15}{100} = \frac{45}{n}$; $n = \$300$
10. part; $\frac{n}{16{,}400} = \frac{20}{100}$; $n = \$3{,}280$

Page 96

1. $\frac{\text{mL}}{\text{lb}}$; $\frac{3}{25} = \frac{n}{125}$; $n = 15$ mL
2. $\frac{\text{lb}}{\$}$; $\frac{40}{120} = \frac{1}{n}$; $n = \$3$
3. $\frac{\text{lb}}{\text{wk}}$; $\frac{3}{1} = \frac{51}{n}$; $n = 17$ wk
4. $\frac{\text{hr}}{\$}$; $\frac{50}{800} = \frac{1}{n}$; $n = \$16$
5. Smith: $\frac{\$}{\text{oz}}$; $\frac{1.19}{16} = \frac{n}{1}$; $n = 7¢$

   Savory: $\frac{\$}{\text{oz}}$; $\frac{0.89}{10} = \frac{n}{1}$; $n = 9¢$

   Smith's costs less.
6. $25.60
7. $6.30

8. \$19.60
9. \$4.50
10. 120 correct

Ratio, Proportion, and Percent Skills Checkup

Pages 97–98

1. C
2. F
3. B
4. J
5. C
6. F
7. B
8. J
9. B
10. J
11. D
12. A
13. F
14. C
15. J
16. B
17. H
18. B
19. F
20. D

Algebra

Page 99

1. 63, 56, 49
2. 27, 33, 39
3. 192, 768, 3,072
4. 12.5, 6.25, 3.125
5. 53, 109, 221
6. 22, 38, 70
7. 5
8. 2
9. 7
10. 3
11. plus 14
12. times 2
13. divided by 5
14. plus 15
15. 32, 48
16. 44, 55
17. 6, 8
18. 10, 11
19. 1,000; 100,000
20. 6.5; 9.5

Page 101

1. $\div$
2. $-$
3. $\times$
4. $-$
5. $\times$
6. $\div, \times$
7. -35
8. -692
9. 2, 8, 10, 12, 48, 60, 60
10. 3, 10, 12, 40, 13, 52, 52
11. 6, 14, 14
12. 15
13. 7
14. 7

Page 102

1. add 4
2. add 6
3. subtract 25
4. multiply by 3
5. 0, 3, 12
6. 5, 9
7. 11, 10
8. 30, 18
9. B

Page 103

1. $16n$
2. $n + 12$
3. $5n$
4. $n + 5$
5. $n - 20$
6. $\frac{1}{8} \times n$ or $\frac{n}{8}$
7. $n + 10$
8. $\frac{n}{2}$ or $n \div 2$
9. $n + 17$
10. $n - 6$
11. $\frac{n}{3}$ or $\frac{1}{3} \times n$
12. $n - 2$

Page 104

1. 7
2. 13
3. 2
4. 20
5. 23
6. 13
7. 26
8. 18
9. 3
10. 80
11. 11
12. 28

Page 105

1. B
2. F
3. B
4. G
5. C
6. F
7. C
8. G

Page 107

1. 18
2. 845
3. 12
4. 88
5. 75
6. 25
7. 240
8. 16
9. 15
10. 12
11. 25
12. 20
13. 3
14. 20
15. 20

16. 8
17. 10
18. 150

Page 108

1. $x < 32$
2. $m < 38$
3. $p > 27$
4. $y < 34$
5. $n < 13$
6. $a < {}^{-}27$
7. $n < 7$
8. $y > 3$
9. $p < 23$
10. $x > 75$
11. $y < 93$
12. $m > 16$

Page 109

1. D
2. F
3. B
4. G
5. B
6. $8 > n$
7. $15 = 3n$
8. $12 < 5y$
9. $\frac{r}{6} = 18$
10. $5n = 75$
11. $14 - n = 2$
12. $n + p = 47$
13. $2y > 8$
14. $\frac{n}{4} > 16$

Page 110

Equations may differ.

1. $(\$2.75 \times n) + \$14 = \$36$; $n = 8$ hours
2. $(\$60 - \$18) \div 2 = n$; $n = \$21$
3. $7n + 5 = 89$; $n = 12$
4. $\frac{n}{2} - \$14.60 = \$3.90$; $n = \$37$
5. $n + \frac{n}{3} = 600$; $n = 450$
6. $6n + 5 = 185$; $n = 30$

Algebra Skills Checkup

Pages 111–112

1. B
2. F
3. A
4. J
5. A
6. H
7. D
8. G
9. D
10. G
11. D
12. H
13. A
14. J
15. B
16. J
17. C
18. G
19. D
20. H

Geometry

Page 113

1. C
2. F
3. B
4. $\cdot C$
5. $\overline{AB}$
6. $\overleftrightarrow{ST}$
7. $\angle MOP$ (O, M, P)

Page 114

1. right
2. acute
3. obtuse
4. right
5. acute
6. obtuse
7. right
8. acute
9. obtuse
10. acute
11. obtuse
12. acute
13. A
14. J

Page 115

1. P
2. I
3. I
4. P
5. yes
6. no
7. no
8. yes
9. $\overleftrightarrow{FE}$
10. no
11. none

Page 117

1. C
2. F
3. C
4. F
5. C
6. G
7. B
8. H
9. B
10. F

Page 118

1. 50°, isosceles and acute
2. 50°, scalene and acute
3. 60°, equilateral and acute
4. 48°, scalene and right
5. A
6. H
7. B
8. H

Page 119

hexagon:
3, 4, 4 × 180° = 720°
octagon:
8, 5, 6, 6 × 180° = 1,080°

1. 108°
2. 120°
3. 135°

Page 120

1. A
2. G
3. line B
4. circumference
5. 2 in.; 4 in.
6. 3 mm, 6mm
7. cannot tell
   cannot tell

Page 121

1. C
2. G
3. B
4. G
5. C and D or A and E
6. & 7. Figures should be drawn correctly.

Page 122

rectangular prism: 6, 12, 8
cube: 6, 12, 8
square pyramid: 5, 8, 5
triangular pyramid: 4, 6, 4
triangular prism: 5, 9, 6

Page 123

1. cylinder
2. cone
3. rectangular prism
4. sphere
5. triangular prism
6. rectangular prism
7. rectangular prism
8. triangular pyramid
9. triangular prism
10. cube

Pages 124–125

1. C
2. F
3. B
4. G
5. A
6. H
7. 90º
8. 360º
9. 180º
10. C
11. J
12. C

Page 126

1. *Y*
2. *B*
3. *N*
4. *E*
5. *S*
6. *H*
7. (2, ⁻2)
8. (4, 6)
9. (1, 0)
10. (2, 7)
11. (5, ⁻3)
12. (0, 3)

Geometry Skills Checkup

Page 127-128

1. A
2. H
3. C
4. J
5. C
6. G
7. B
8. F
9. H
10. B
11. J
12. A
13. J
14. D
15. H
16. C
17. F

Measurement

Page 129

1. 19
2. 95
3. 70
4. 65
5. 25
6. 70
7. 50
8. 75

Page 130

9. 2
10. 2
11. 5
12. 20
13. 20
14. 4
15. 25, 75
16. 2, 32
17. 10, 20
18. 1, 43

Page 131

1. $1\frac{1}{2}$
2. 2
3. $\frac{1}{2}$
4. $2\frac{1}{2}$

5. 1
6. $1\frac{1}{2}$
7. $\frac{1}{2}$
8. $\frac{1}{4}$
9. $\frac{3}{4}$
10. $1\frac{1}{4}$
11. $\frac{1}{2}$
12. $1\frac{3}{4}$
13. 1
14. 2

Page 132

1. 4 by $7\frac{1}{2}$ in.
2. $\frac{1}{4}$ by $7\frac{1}{2}$ in.
3. 3 by 5 in.
4. 8 by 11 in.
5. $2\frac{1}{2}$ by $4\frac{1}{2}$ in.
6. 4 by 12 in.
7. $2\frac{5}{8}$ by 4 in.
8. $3\frac{1}{2}$ by $3\frac{5}{8}$ in.
9. $2\frac{5}{8}$ by $6\frac{1}{8}$ in.
10. 1 by $1\frac{3}{4}$ in.
11. A
12. H
13. D
14. J
15. B
16. H

Page 133

1. C
2. G
3. B
4. F
5. 15 inches
6. 2 yards
7. 1 pound
8. 22 ounces
9. 3 cups
10. 1 quart
11. 1 gallon
12. 2 hours

Page 134

1. 3
2. 2
3. 1
4. 6
5. 1
6. 18
7. 5
8. 48
9. 8
10. 2
11. 4,000
12. 6
13. 90
14. 15
15. 14
16. 3, 2
17. 1, 7

Page 135

1. C
2. F
3. G
4. C
5. E
6. H

Page 136

1. 500
2. 2,000
3. 3
4. 1,500
5. 3.5
6. 100
7. 2,500
8. 1.5
9. 315
10. 1,450
11. 3,015
12. 1 kg 515 g

Page 137

1. 1 in.
2. 1 oz
3. 1 L
4. 1 m
5. 1 mi
6. 18.2
7. 1.4
8. 25
9. 11
10. 3.3
11. 6.8
12. 40
13. 0.5
14. 30
15. 2.4

Page 138

1. 4 hr
2. 2 lb 1 oz
3. 8 yd 1 ft
4. 16 ft 5 in.
5. 2 hr 5 min
6. 3 qt 1 c
7. 7 lb 3 oz
8. 1 yd 1 ft
9. 3 ft 8 in.
10. 2 qt
11. 2 hr 30 min
12. 2 pt 1 c

Page 139

1. 10:15
2. 8:45 A.M.
3. 4:45
4. 40 min
5. 9:10

Page 140

1. 1 hr 22 min
2. 2 hr 25 min
3. 1 hr 40 min
4. 4 hr 45 min
5. 4 hr 15 min
6. 2 hr 50 min
7. 2 hr 9 min
8. 6 hr 30 min

Page 141

1. 60 ft
2. 18 cm
3. 24 in.
4. 22 ft 10 in.
5. 48 m
6. 16 mm
7. 32 km
8. 4 in.
9. 55 ft
10. 28 ft
11. 4 in.
12. 18 cm

Page 142

1. 31.4 in.
2. 37.68 m
3. 75.36 yd
4. 22 ft
5. 44 in.
6. 66 cm

Page 143

1. 84 mi^2
2. 25m^2
3. 90 $in.^2$
4. 180 ft^2
5. 6 m^2
6. 6 yd^2

Page 144

1. 10 $units^2$
2. $4\frac{1}{2}$ $units^2$
3. 6 $units^2$
4. 14 $units^2$
5. 6 $units^2$
6. 35 $units^2$

Page 145

1. 4 ft, 8 ft; 16 ft^2
2. 15 m, 5 m, 37.5 m^2
3. 20 cm, 24 cm; 240 cm^2
4. 25.9 yd^2
5. $1\frac{1}{2}$ ft^2
6. 18.3 m^2
7. 10.9 $in.^2$
8. 16.4 m^2
9. 10.7 cm^2

Page 146

1. 6 m, 36 m; 113.04 m^2
2. 20 ft, 400 ft; 1,256 ft^2
3. 5 yd, 25 yd; 78.5 yd^2
4. 2 km, 4 km: 12.56 km^2
5. 14 in., 196 in; 616 $in.^2$
6. 7 m, 49 m; 154 m^2

Page 147

1. 93 cm^2
2. 150 yd^2
3. 440 ft^2
4. 232 $in.^2$

Page 148

1. 4 cm^3
2. 4 cm^3
3. 4 cm^3
4. 8 cm^3
5. 6 cm^3
6. 27 cm^3
7. 12 cm^3
8. 7 cm^3
9. 12 cm^3

Measurement Skills Checkup

Pages 149–150

1. C
2. G
3. B
4. J
5. C
6. F
7. D
8. F
9. C
10. G
11. D
12. F
13. C
14. H
15. B

Probability, Data, and Statistics

Page 152

1. D
2. H
3. B
4. G
5. C
6. J
7. D
8. F

Page 153

1. D
2. J
3. C

Page 154

1. B
2. Nov. 21, 1980
3. 30,190,300
4. M*A*S*H*, Dallas, and Roots
5. M*A*S*H*
6. Jan. 30, 1977
7. M*A*S*H*

Page 155

1. $4.65
2. $2.28
3. $10.38
4. $12.94
5. $19
6. none
7. $3.25

Page 156

1. Cherokee
2. Choctaw
3. 4,894
4. Sioux
5. Navajo
6. 29,360
7. Sioux
8. D

Page 157

1. A
2. H
3. C
4. H
5. C

Page 158

Ben Hanks	180	no mode	180
Jose Ruiz	230	230	227
Phil Chu	173.5	175	172
Linda Glass	165	no mode	165
Tim Horne	121.5	144	121
Connie Chu	229	246	214

Page 159

1. C
2. F
3. D
4. B
5. H
6. A

Page 160

1. 12,000
2. 10,000
3. 4,000
4. 8,000
5. Delays/Cancellations and Customer Service
6. 2,000

Page 161

1. 500 million
2. D
3. twice as many
4. C

Page 162

1. 200
2. $\frac{550}{100{,}000} \approx \frac{11}{200}$
3. cancer
4. cancer and diabetes
5. cancer and diabetes
6. $\frac{300}{100{,}000} = \frac{3}{1{,}000}$
7. 400
8. heart disease
9. heart disease

Page 163

1. A
2. F
3. C
4. H
5. D
6. J

Probability, Data, and Statistics Skills Checkup

Pages 164–165

1. B
2. F
3. A
4. J
5. A
6. G
7. B
8. F
9. B
10. J
11. A
12. F
13. C
14. J

Glossary

absolute value (| |) a number's distance from 0 on the number line. Both 5 and −5 have an absolute value of 5.

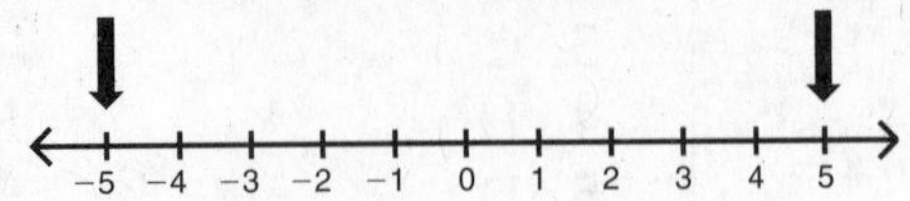

acute angle an angle that measures less than 90°

angle (∠) the figure formed by two rays meeting at a common endpoint

area the size of a flat region, generally expressed in square units

average *See* mean

canceling a method of simplifying fractions before multiplying them. To cancel, divide a number in the numerator and a number in the denominator by the same factor.

circumference the distance around the outside of a circle, found by multiplying the circle's diameter by π (approximately 3.14)

common denominator a whole number that is a denominator of all members of a group of fractions. The **least common denominator** (LCD) is the smallest number that can be evenly divided by all denominators of the group.

congruent (≅) having the same size and shape

coordinate plane a flat surface with two number lines called axes that intersect at a right angle

coordinates an ordered pair of numbers that identify the location of a point on the coordinate plane. The first number of the pair represents the horizontal distance from the origin (0, 0) along the x-axis, and the second number represents the vertical distance from the origin along the y-axis.

cross product the product of the numerator of one fraction or ratio and the denominator of a second fraction or ratio. If two fractions are equal, their cross products will be equal. Example: If $\frac{3}{10} = \frac{b}{15}$, then $3 \times 15 = b \times 10$.

decimal a number that uses a decimal point and place value to show values less than 1 Examples: 0.5 and 1.67

decimal point a period used to separate a whole number from an amount less than 1

denominator the bottom number of a fraction. The denominator tells how many equal parts there are in one unit.

edge the line along which two faces of a solid meet

equation a number sentence that uses an equal sign to indicate that two amounts are equal

equivalent having the same value. Examples: $\frac{1}{2}$, $\frac{2}{4}$, and $\frac{5}{10}$ are equivalent fractions.

expression a part of a number sentence that combines numbers, variables, and operation signs. Example: $3 + n$

face a flat surface of a three-dimensional figure

factor a number that is multiplied with another number

fraction a number written in the form $\frac{a}{b}$ that represents part of a whole

function a rule that changes one number to another number

improper fraction a fraction in which the numerator is greater than the denominator

inequality a mathematical sentence that shows that one value does not equal another Examples: $3 > 5$ and $5 \neq 7$

integers positive whole numbers and their opposites, together with 0

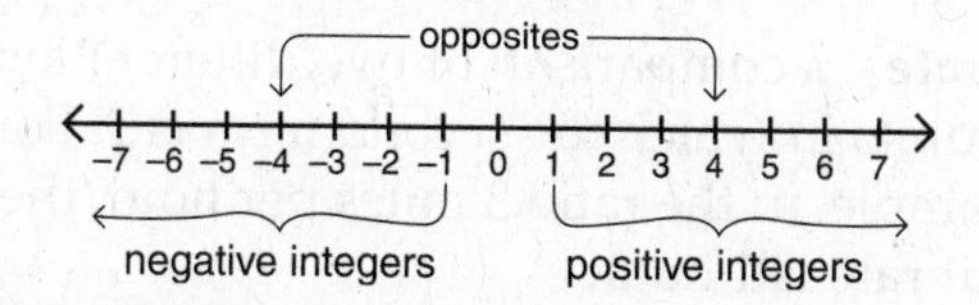

inverse operations operations that are opposites and undo each other. Addition and subtraction are inverse operations. For example, if you add 5, you can undo the addition by subtracting 5. Multiplication and division are also inverse operations.

key an explanation of symbols used in a map, graph, diagram, or table. Also called a legend.

line segment part of a line. A line segment has two endpoints.

mean the sum of a set of values divided by the number of values in the set

median the middle value in a set of numbers listed in order

mixed number the combination of a whole number and a fraction. A mixed number represents a value between two whole numbers.

mode the number in a set of data that appears most often

multiple the product of a given number and a whole number. 3, 6, 9, 12,… are multiples of 3.

negative number a number with a value less than 0. A negative number is shown with a negative sign. For example, negative 3 would be written −3.

numerator the top number of a fraction. The numerator tells how many equal parts of the whole are represented.

obtuse angle an angle that measures more than 90° but less than 180°

ordered pair two numbers that name the x-coordinate and y-coordinate of a point

outcome a possible result in a probability experiment

parallel lines (||) straight lines that never meet

percent (%) a number expressed in relation to 100 with a percent sign (%). Example: If 100 people voted, and 14 of them voted for Smith, then Smith got 14% of the vote.

perimeter the total length around the outside of a closed figure

perpendicular lines (⊥) straight lines that intersect (or will intersect) at a 90° angle

pi (π) the ratio of the circumference of a circle to its diameter, approximately 3.14 or $\frac{22}{7}$

point a location of an object or position in space

polygon a simple flat shape that is closed and has straight sides. Examples: triangles, squares, and rectangles.

probability the likelihood or chance of an event occurring

proportion a statement showing that one ratio is equal to another

quadrilateral a four-sided polygon

radius the distance from the center of a circle to its circumference

ratio a comparison of two numbers

ray a part of a line that extends in one direction

reflection (flip) a mirror image that is the result of a figure being flipped over a line

regular polygon a polygon with congruent sides and congruent angles

right angle an angle that measures 90°

rotation (turn) the result of a turn of a figure around a fixed point

similar (~) having the same shape, but not necessarily the same size

simplest terms a fraction or ratio in which the only factor common to the numerator and denominator is 1

surface area the sum of the areas of all the faces of a solid figure

three-dimensional figure a figure that has length, width, and depth. Also called a solid. Examples: cones, cubes, and cylinders

transformation the movement of a figure by reflection, rotation, or translation

translation (slide) a move to another location without rotating or reflecting the figure

unit rate a comparison of two different kinds of units in which the second term is 1. For example, in the ratio 3 miles per hour, the unit rate is 1 hour.

variable a letter or shape that represents a number in an expression or equation. Also called an unknown.

vertex the point at which lines meet. Also, a corner, or a point at which three or more faces of a solid meet.

volume the amount of space occupied by a three-dimensional figure measured in cubic units

Index